ENGINE AND PARTS MANUALS FOR THE MEP 804/814 A/B GENERATOR SETS

TM 9-2815-254-24 (ENGINE)
TM 9-2815-254-24P (PARTS)

GENERATOR SET SKID MOUNTED
TACTICAL QUIET
15 KW, 50/60 HZ, MEP-804A
(NSN: 6115-01-274-7388) (EIC: VG4)
15 KW, 50/60 HZ, MEP-804B
(NSN: 6115-01-530-1458) (EIC: N/A)
AND
GENERATOR SET SKID MOUNTED
TACTICAL QUIET
15 KW, 400 HZ, MEP-814A
(NSN: 6115-01-274-7393) (EIC: VN4)
15 KW, 400 HZ, MEP-814B
(NSN: 6115-01-529-9494) (EIC: N/A)

edited by
Brian Greul

This is the Engine and Parts manuals for the MEP series of Military Generators are reknowned for their quiet, durable operation and conservative power ratings.   This is the operators manual for the 15KW version of the generator issued under models 804 and 814. The A series has analog controls and the B series has digital controls. Various units are manufactured for the US Government by different contractors with different power plants. This book is a reprint of the operator manual published by the US Army. It is printed as a courtesy to enthusiasts and owners of these generator sets. Other important manuals for this generator are also printed by this publisher.

An 8.5x11 3 hole punched loose leaf copy may be purchased for your 3 ring binder. Email books@ocotillopress.com for current information.

Should you have suggestions or feedback on ways to improve this book please send email to Books@OcotilloPress.com We also welcome requests for other military manuals you would like to see printed.

Edited 2021 Ocotillo Press
ISBN 978-1-954285-18-7

Printed in the United States of America

Ocotillo Press
Houston, TX 77017
Books@OcotilloPress.com

TECHNICAL MANUAL

UNIT, DIRECT SUPPORT AND GENERAL SUPPORT MAINTENANCE INSTRUCTIONS

DIESEL ENGINE MODEL C-240PW-28 4 CYLINDER 2.4 LITER NSN: 2815-01-350-2207

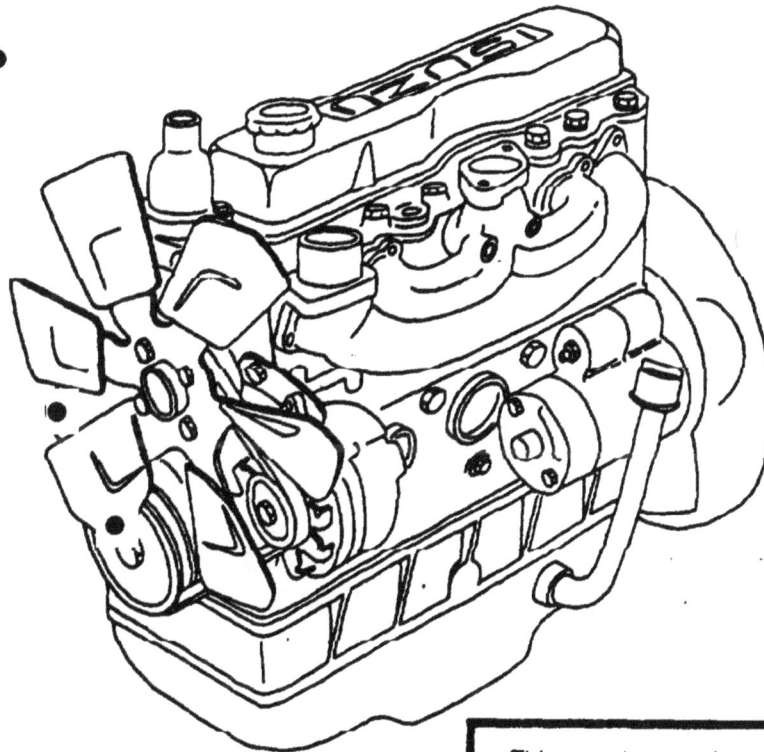

This copy is a reprint which includes current pages from Changes 1 and 2.

HEADQUARTERS, DEPARTMENTS OF THE ARMY AND THE AIR FORCE 1 SEPTEMBER 1993

A

SAFETY SUMMARY

For first aid, refer to FM21-11.

WARNING

The noise level when operating could cause hearing damage. Ear pro-

WARNING

tection must be worn.

WARNING

Where applicable, prior to performing engine maintenance, ensure bat-teries are disconnected.
Do not drain coolant until the coolant temperature is below operating

WARNING

temperature. Always loosen cooling system filler cap, radiator cap, or drain cock slowly to relieve any excess pressure.
Diesel fuel is flammable and toxic to eyes, skin, and respiratory tract.

WARNING

Skin/eye protection required. Avoid repeated/prolonged contact. Good general ventilation is normally adequate.

WARNING

Cleaning solvent is flammable and toxic to eyes, skin, and respiratory tract. Skin/eye protection required. Avoid repeated/prolonged contact. Good general ventilation is normally adequate.

CHANGE

No.2

HEADQUARTERS, DEPARTMENTS
OF THE ARMY AND THE AIR FORCE
WASHINGTON, D.C., 30 October 1996

Unit, Direct Support and General Support
Maintenance Instructions

DIESEL ENGINE
MODEL C-240PW-28 4
CYLINDER 2.4 LITER NSN:
2815-01-350-2207

DISTRIBUTION STATEMENT A: Approved for public release; distribution is unlimited

TM 9-2815-254-24/TO 38G1-94-2, 1 September 1993, is changed as follows:

1. Remove and insert pages as indicated below. New or changed text material is indicated by a vertical bar in the margin. An illustration change is indicated by a miniature pointing hand.

Remove pages i
and ii
3-1 and 3-2
3-35 and 3-36
3-69 through 3-76
3-79 through 3-82
3-85 through 3-86
3-91 through 3-96
3-147 through 3-150

3-163 through 3-166
Insert pages i
and ii
3-1 and 3-2
3-35, and 3-36
3-69 through 3-76
3-79 through 3-82
3-85 through 3-88
3-91 through 3-96
3-147 through 3-150

2. Retain this sheet in front of manual for reference purposes.

By Order of the Secretaries of the Army and Air Force:

DENNIS J. REIMER
General, United States Army
Chief of Staff

Official:

JOEL B. HUDSON
Administrative Assistant to the
Secretary of the Army
02954

RONALD R. FOGELMAN
General, USAF
Chief of Staff

HENRY VICCELLIO Jr.
General, USAF
Commander, Air Force Material Command

DISTRIBUTION:
    To be distributed in accordance with DA Form 12-25-E, block no. 5143, requirements for TM 9-2815-254-24.

CHANGE

No. 1

HEADQUARTERS
DEPARTMENTS OF THE ARMY, AND AIR FORCE
WASHINGTON, DC., 15 DECEMBER 1993

TECHNICAL MANUAL

UNIT, DIRECT SUPPORT AND GENERAL SUPPORT MAINTENANCE INSTRUCTIONS DIESEL ENGINE MODEL C-240PW-28 4 CYLINDER 2.4 LITER NSN: 2815-01-350-2207

TM 9-2815-254-24/TO 39G1-94-2, 1 SEPTEMBER 1993 is changed as follows:

1. Remove and insert pages as indicated below. New or changed text material is indicated by a vertical bar in the margin. An illustration change is indicated by a miniature pointing hand.

Remove pages	Insert pages
iii and iv	iii and iiv 3-3and 3-4
3-3 and 3-4 3-9through 3-12	3-9 through 3-12 B-5
and B-6	B-5 and B-6
------Appendix E	

2. Retain this sheet in front of manual for reference purposes.

By Order of the Secretaries of the Army and air Force

GORDON R. SULLIVAN
General, United States Army
Chief of Staff

Official:

Milton H. Hamilton

MILTON H. HAMILTON
Administrative Assistant to the
Secretary of the Army
05829

MERRILL A. McPEAK
General, USAF
Chief of staff

Official:

RONALD W. YATES
General, USAF
Commander, Air Force Material Command

DISTRIBUTION:
To be distributed in accordance with DA Fom 12-25-E, block no. 5143, requirements for TM 9-2815-254-24.

TECHNICAL MANUAL

NO. 9-2815-254-24

HEADQUARTERS
DEPARTMENTS OF THE ARMY AND THE AIR FORCE
WASHINGTON, D.C., 1 September 1993

Unit, Direct Support and General Support Maintenance Instructions

**DIESEL ENGINE
MODEL C-24OPW-28 4
CYLINDER 2.4 LITER NSN:
2815-01-350-2207**

REPORTING ERRORS AND RECOMMENDING IMPROVEMENTS

You can help improve this manual.If you find any mistakes or if you know of a way to improve these procedures, please let us know.

(A) Mail your letter or DA Form 2028 (Recommended Changes to Publications and Blank Forms), or
DA Form 2028-2 located in the back of this manual directly to: Commander, US Army Aviation and Troop
Command, ATTN: AMSAT-I-MP, 4300 Goodfellow Blvd.,St. Louis, MO 63120-1798. You may alsosubmit your
recommended changesby E-mail directly to< mpmt%avma28@st-louis-emh7.army.mil>.Instructions for sending
an electronic 2028 may be found at the back of this publication immediately preceding the hard copy 2028.

(F) Air Force - AFTO Form 22 Directly to:Commander, Sacramento Air Logistics Center, ATTN:
TILBA, McClellan AFB, CA 95652-5990 (AFMC)

A reply will be furnished directly to you.

TABLE OF CONTENTS

TABLE OF CONTENTS - Continued

TABLE OF CONTENTS - Continued

LIST OF ILLUSTRATIONS

LIST OF ILLUSTRATIONS - Continued

LIST OF ILLUSTRATIONS - Continued

LIST OF TABLES

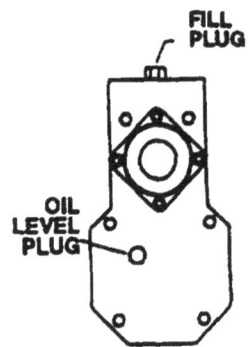

FILL
PLUG

OIL
LEVEL
PLUG

CHAPTER 1
INTRODUCTION
SECTION I. GENERAL INFORMATION

1-1. **SCOPE.**

1-1.1. Type of Manual. This manual contains unit, direct support, and general support maintenance instructions for the Model C240 Diesel Engine, hereafter referred to as engine. Also included are descriptions of major systems/com-ponents and their functions in relation to other systems/components.

1-1.2. Purpose of Equipment. The engine provides a driving force for generators or other equipment requiring this size (HP rating) and compatibility.

1-2. MAINTENANCE FORMS RECORDS AND REPORTS.

1-2.1. Reports of Maintenance Unsatisfactory Equipment. Department of the Army forms and procedures used for equip-ment maintenance will be those prescribed by DA Pam 738-750, the Army Maintenance Management Sys-tem (TAMMS). Air Force personnel will use AFR 66-1, Air Force Maintenance Management Policy, for mainte-nance reporting and TO-00-35D54 for unsatisfactory equipment reporting.

1-2.2. Reporting of Item and Packaging Discrepancies. Fill out and forward SF 364 (Report of Discrepancy (ROD)) as prescribed in AR 735-11-2/DLAR 414-55/SECNAVINST 4355.18/AFR 400-54/MCO 4430.3J.

1-2.3. Transportation Discrepancy Report (TDR) (SF 361). Fill out and forward Transportation Discrepancy Report (TDR) (SF 331) as prescribed in AR 55-38/NAVSUPINST 4610.33C/AFR 75-18/MCO P4610.19D/DLAR 4500.15.

REPORTING EQUIPMENT IMPROVEMENT RECOMMENDATIONS (EIR).

1-3.
1-3.1. Army. If your Military Standard Engine needs improvement, let us know. Send us an EIR. You, the user, are the only one who car tell us what you don't like about your equipment. Let us know why you don't like the design or performance. Put it on an SF 368 (Product Quality Deficiency Report). Mail it to us at: Commander, US. Army Avi-ation and Troop Command, ATTN: AMSAT-I-MDO, 4300 Goodfellow Boulevard, St. Louis, Missouri 63120-1798. We will send you a reply.
1-3.2. Air Force. Air Force personnel are encouraged to submit EIR's in accordance with AFR 900-4.

1-4.   DESTRUCTION OF ARMY MATERIAL TO PREVENT USE

Refer to TM 750-244-3 for procedures to destroy equipment to prevent enemy use.

1-5.   **PREPARATION FOR STORAGE OR SHIPMENT.**

Refer to TB 740-97-2 for procedures to place the equipment into storage.

1-6.   WARRANTY.

The engine is warranted for a specific period of time. Refer to Warranty Technical Bulletin for the end item. The warranty starts on the date found in block 23, DA Form 2408-9, in the equipment log book. Report all defects in material or workmanship to your supervisor, who will take appropriate action.

SECTION II. EQUIPMENT DESCRIPTION AND DATA

1-7. GENERAL.

The diesel engine (FIGURE 1-1) is four cylinder, four cycle, fuel injected, naturally-aspirated, and liquid-cooled. The firing order is 1-3-4-2. The number one cylinder is toward the fan end of the engine. The serial number is found on left side of the cylinder body at number one cylinder location. Rotation of engine is counterclockwise as viewed from flywheel.

NOTE

All locations referenced herein are given facing the flywheel end (rear) of the engine.

1-8.   DETAILED DESCRIPTION.

1-8.1. Cooling System. The cooling system consists of a radiator, water pump, cooling fan, thermostat, and connecting hoses. The fan is mounted on shaft of water pump and both are belt driven from the crankshaft pulley. The thermo-stat controls engine temperature and is installed in the top of engine. The function of the cooling system is to main-tain a specific operating temperature of 180° to 203°F (82° to 95°C) for the engine.

1-8.2. Lubrication System. The lubrication system consists of the oil pan, a gerotor type oil pump, spin-on type oil filter, and internal passages within the engine. There are two external oil lines, one for rocker arm lubrication and one for fuel injector pump lubrication.

1-8.3. Fuel System. The function of the fuel system is to inject a metered quantity of clean atomized fuel into the engine cylinders at a precise time near the end of the compression stroke of each piston. The fuel system consists of the fuel tank, fuel filter/water separator, fuel injection pump, and the fuel injectors. The fuel tank and fuel filter/wa-ter separator are not mounted on engine. The injection pump is mounted on the front cover and is driven by the timing gears.

1-8.4. Electrical System. The electrical system is 24 VDC operation and consists of an alternator, starter, externally mounted batter and other items as required. The alternator is mounted on front of engine and is belt driven. When the engine is operating, the alternator supplies 24 VDC to recharge the battery and maintain it at a full state of charge. The starter is mounted on the flywheel housing and when energized engages the ring gear of the fly-wheel to rotate the engine.

1. Fan
2. Fan Belt
3. Fan Pulley
4. Fuel Pipes
5. Fuel Injector Pipes
6. Injector Nozzle and Glow Plug
7. Oil Filter
8. Fuel Filter Housing
9. PCV Valve
10. Crankshaft Pulley
11. Battery Charging Alternator
12. Starter
13. Intake Manifold
14. Exhaust Manifold
15. Valve cover
16. Thermostat Housing
17. Water Pump
18. Oil Pan
19. Flywheel Housing
20. Crankcase
21. Fuel Injection Pump
22. Dipstick and Tube

RIGHT SIDE

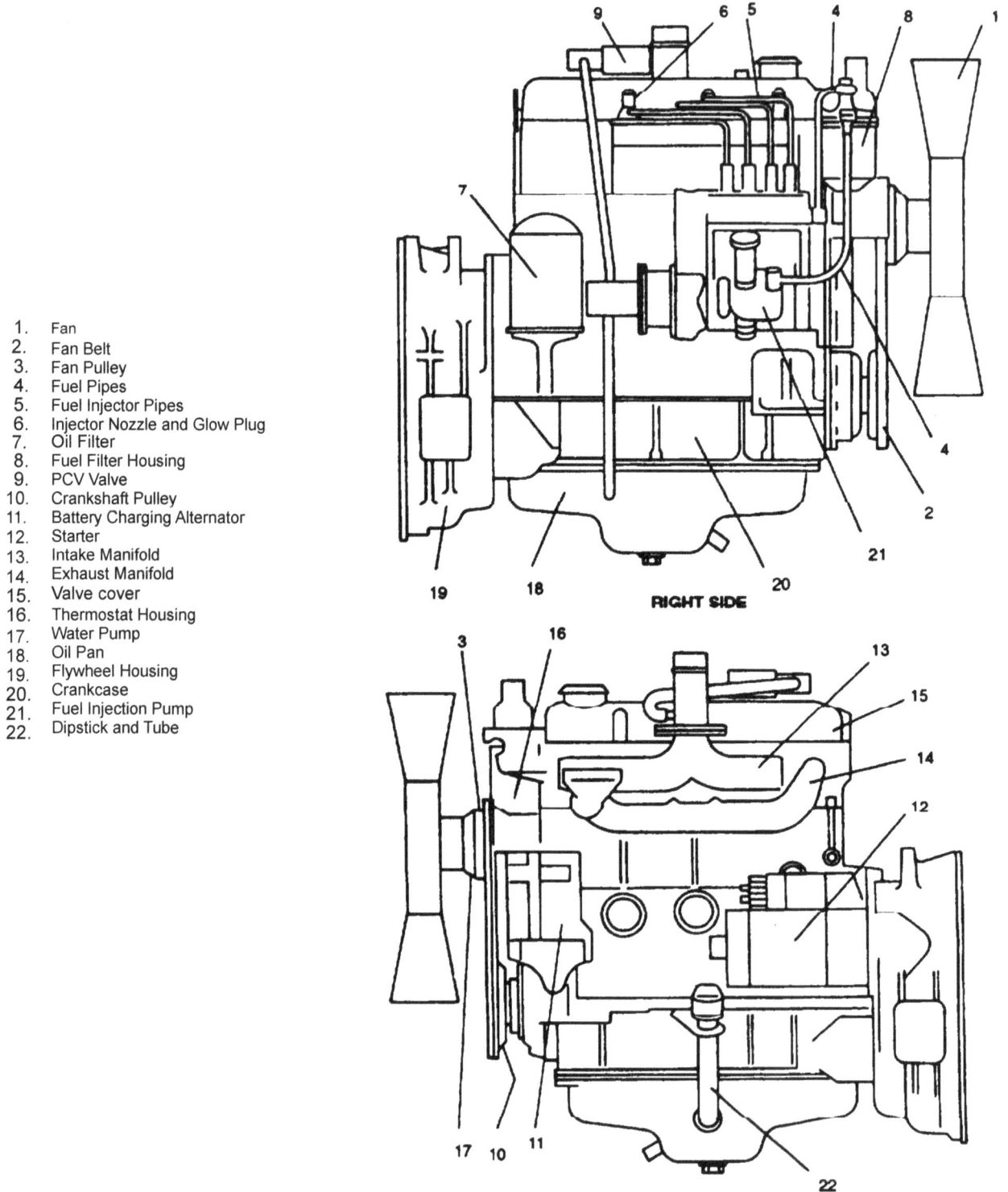

FIGURE 1-1. Engine Components

1-9. EQUIPMENT DATA.

1-9.1. Leading Particulars. For a list of leading particulars, refer to TABLE 1-1.

TABLE 1-1. Table of Specifications

Model...Isuzu, C240 PW-28
Type... Four cylinder, four cycle, liquid cooled diesel
Bore/Stroke..3.38/4.01 in. (86/102 mm)

Displacement · 144.5 cu in. (2369 cu cm)
Horsepower Rating..
Compression Ratio..
20:1 Length· 28.9 in (736 mm)
Width ·· 21.0 in (535 mm)
Height · 32.1 in (815 mm)
Weight · 491 lbs (223 kg)
Firing Order · 1-3-4-2
Injection Pump ·· Kiki - PES4A
Governor ....Barber-Colman Model 10502-001-0-2
Injection Starting Pressure...1706 psi (11,772 kPa)
Cylinder Compression Pressure (warm)..............................441 psi (3041 kPa)
Valve Clearance (cold) ···0.017 in. (0.45 mm)
Lubrication System Capacity...1.6 gal (6.1 liters)
Coolant System Capacity (engine only) ·· 1.37 gal (5.2 liters)
Battery Charging Alternator... Hitachi, 24 VDC - 20 amp
Starter..Hitachi,24VDC-3.5kw

SECTION III. PREPARATION FOR USE

1-10. INSPECTING AND SERVICING ENGINE.

This section provides information and guidance for inspecting, servicing, and installing the engine. For additional information, also refer to end item maintenance manual.

1-10.1. Inspection.

    a. Check that all packing materials have been removed.

    b. Check engine identification plate for positive identification.

    c. Inspect engine exterior for shipping damage.

    d. Check fan belt drive for proper tension. Refer to end item maintenance manual.

    e. Inspect engine for loose or missing mounting hardware, or damaged or missing parts.

1-10.2. Service. Except for servicing the lubrication system all other servicing must be accomplished after engine is mounted in the end item, refer to end item maintenance manual and lubrication order (LO).

CHAPTER 2
OPERATION
SECTION I. PRINCIPLES OF OPERATION

2-1.   INTRODUCTIONS

This section contains functional descriptions of the engine systems and how they are connected to the end item

2-2.   COOLING SYSTEM.

The cooling system consists of a radiator, hoses, thermostat, belt driven fan, water pump, and cooling jackets within the engine. The water pump forces coolant through passages (coolant jackets) in the engine block and cylinder head where coolant absorbs heat from the engine. When the coolant temperature is below operating tem-perature, the thermostat is closed and coolant is bypassed to the water pump inlet. As coolant temperature in-creases to 180°F (82°C), the thermostat starts to open, restricting bypass flow and opening flow to the radiator. As coolant temperature continues to increase to 203°F (95°C), the thermostat is fully opened, shutting off all bypass flow and providing full flow through the radiator. Air forced through the fins of the radiator by the fan cools the coolant pumped through the radiator. Items are added to the engine to monitor coolant temperature and to warn if temperature exceeds a predetermined value.

2-3.   LUBRICATION SYSTEM.

The lubrication system consists of an oil pan, dipstick, pump, and filter. The oil sump is a reservoir for lubricating oil. The dipstick indicates oil level in sump. The pump draws oil from the sump through a screen which removes large impurities. The oil then passes through a spin-on type filter where small impurities are removed. From the filter, oil enters the cylinder head oil gallery and is distributed to the engines internal moving parts and to the fuel injection pump. After passing through the block and pump, the oil returns to the oil sump. Items are added to moni-tor oil pressure and to warn/stop engine if pressure drops to a dangerously low value.

2-4. FUEL SYSTEM.

2-4.1. The fuel system consists of an external fuel tank, transfer pump, fuel filter/water separator, fuel injection pump, fuel injectors, and piping. Fuel from an external source is supplied to the fuel injection pump. The injection pump provides a pressurized metered quantity of clean atomized fuel through the injector nozzles into the cylinder at a precise time near the end of the compression stroke of each piston. The fuel that is not used by the injectors is returned to the fuel tank via an excess fuel return line.

2-4.2. Extremely cold outside temperatures make starting the engine difficult. To improve engine starting, a cold weather starting aid has been provided that features a glow plug for each cylinder. The glow plugs are energized to preheat engine combustion air during engine preheat starting cycle.

2-5.   <u>ELECTRICAL SYSTEM.</u>

The electrical system consists of external mounted batteries, starter, battery charging alternator, and related re-lays and switches for control of the system. Battery power supplied to the starter during the start cycle energizes the starter which engages the ring gear of the flywheel causing the engine to turn over. When engine start is com-plete, the starter is deenergized and disengages from the flywheel. The battery charging alternator is belt driven. It is a 20 ampere, 24 VDC alternator that when operating, supplies voltage to the 24 VDC circuit and recharges the batteries and maintains them at a full state of charge.

SECTION II. OPERATING INSTRUCTIONS

NOTE

Refer to end item operator's manual.

CHAPTER 3

MAINTENANCE

Section I. PREVENTIVE MAINTENANCE CHECKS AND SERVICES (PMCS)

3-1. PMCS PROCEDURES.

3-1.1. General.

To ensure that engine is ready for operation at all times, it must be inspected so defects can be discovered and corrected before they result in serious damage or failure. Perform operator's PMCS prior to or in conjunction with performance of engine PMCS. For engine PMCS, refer to TABLE 3-1.

Table 3-1. Preventive Maintenance Checks and Services (PMCS)

Item No.	Interval						Item to be Inspected	Procedures Check for and have repaired or adjusted as necessary	Equipment is Not/Ready Available If:
	M	Q	S	A	B	H			
1						300	Oil Filter	Change oil filter. Refer to paragraph 3-14.1. NOTE Oil filter change interval is a hard timereplacement to beused when AOAPlab is not available.	
2						750	PCV System	Service PCV system. Refer to paragraph 3-20.	
3						1500	Cylinder Head Bolts	Torque cylinder head bolts. Refer topara-graph 3-27.6d.	
4						3000	Engine Valve Clearances	NOTE Valve clearancemust be checkedafter cylinder headbolt tightening hasbeen done. Refer tostep 3 in this table. Adjust engine valve-clearances. Refer toparagraph 3-26.6.	

Table 3-1. Preventive Maintenance Checks and Services (PMCS) - Continued

Item No.	Interval						Item to be Inspected	Procedures Check for and have repaired or adjusted as necessary	Equipment is Not/Ready Available If:
	M	Q	S	A	B	H			
5							3000 Engine Fuel Injectors	Remove, clean, andtest injectors. Referto paragraph 3-10.	
6							1500 Engine Compression	Check engine com-pression. Refer toparagraph 3-27.7.	Engine compressionis low.
7							1500 Engine Oil Pressure	Check engine oilpres-sure. Refer toparagraph 3-19.	Engine oil pressurenot as specified.
6						300	Fuel Injection Pump Governor Oil	Remove oil levelplug to check oillevel. Fill at fill plug. Change oil if itbecomes con-taminated.	

SECTION II. TROUBLESHOOTING

3-2.   TROUBLESHOOTING PROCEDURES.

3-2.1. Purpose of Troubleshooting Table. This section contains troubleshooting information for locating and Correcting operating troubles which may develop in the engine. Each malfunction for an individual component unit or system is followed by a list of tests or inspections which will help you to determine probable causes and corrective action to take. You should perform tests/inspections and corrective actions in order listed.

This table cannot list all malfunctions that may occur, nor all tests or inspections and corrective actions. If a mal-function is not listed or cannot be corrected by listed corrective actions. notify your supervisor.

NOTE

Before you use this table, be sure you have performed your PMCS. Prior to performing troubleshooting procedures within this manual, per-form your operator's troubleshooting and the end item maintenance manual troubleshooting procedures.

SYMPTOM INDEX

TABLE 3-2 Troubleshooting

MALFUNCTION
  TEST OR INSPECTION
    CORRECTIVE ACTION

---

1.   ENGINE WILL NOT CRANK

   Step 1. Check for defective end item starting system.
       Troubleshooting per end item maintenance manual. If not defective, do step 2.

   Step 2. Check for defective starter motor and solenoid.
       a Test starter and solenoid. Refer to paragraph NO TAG

       b. Repair/replace defective starter and/or solenoid. Refer to paragraph 3-23.

---

2.   STARTER OPERATES BUT ENGINE DOES NOT TURN OVER.

   Step 1. Check for worn or broken starter pinion gear and/or flywheel ring gear.

       a. Remove starter and inspect pinion gear and flywheel ring gear for damage.

       b.Replace defective clutch assembly and/or flywheel ring gear. Refer to paragraphs 3-23 and 3-31.

   Step 2. Crankshaft rotation restricted. Attempt to manually rotate engine. If unable to manually rotate, repair/replace engine.

---

3.   ENGINE CRANKS BUT FAILS TO START

   Step 1. Check for fuel being supplied to fuel injection pump.

       a. Test feed pump capacity. Refer to paragraph NO TAG. If feed pump not defective, do step 2.

       b. Repair or replace fuel feed pump. Refer to paragraph 3-11. Check

   Step 2. for dogged fuel filter/water separator.

       a Remove and dean fuel filter Assembly, paragraph 3-9. If dean, replace water separator, refer to end item maintenance manual.

   Step 3. Check for air in fuel system lines.

       Bleed fuel system. Refer to paragraph 3-12. If fuel system is free air, do step 4.

   Step 4. Check for fuel injector starling pressure too low or spray condition improper.

       a Remove, dean and test fuel injector. Refer to paragraph 3-10. Adjust as necessary, if injector can not be adjusted, go to b. If not defective, do step 5.

       b. Replace fuel injector. Refer to paragraph 3-10. Check

   Step 5. for improper fuel inject pump timing.

       c. Check fuel injection pump timing. Refer to paragraph 3-13.1. If fuel injection pump timing is correct, do step 6.

       b. Adjust fuel injection pump timing. Refer to paragraph 3-132. Check

   Step 6. for defective fuel injection pump.

       c. Remove and test fuel injection pump. Refer to paragraph NO TAG.

       b. Repair or replace defective fuel injection pump. Refer to paragraph 3-11.

Table 3-2. Troubleshooting - Continued

MALFUNCTION
TEST OR INSPECTION
CORRECTIVE ACTION

4._____ <u>ENGINE HARD TO START OR WILL NOT START IN COLD WEATHER.</u>

 step 1. Check for faulty glow plugs.

 a. Test glow plugs. Refer to paragraph 3-22.3. If glow plugs not defective, do step 2.

 b. Replace defective glow plugs. Refer to paragraph 3-22. Step 2.

Refer to Malfunction 3 and perform steps 1 through 5.

Step 1. Check for fuel being supplied to fuel injection pump.

 a. Test feed pump capacity. Refer to paragraph 3-11.5 step aj. If feed pump not defective, do step 2.

 b. Repair or replace fuel feed pump. Refer to paragraph 3-11. Check

Step 2. for air in fuel system lines.

Bleed fuel system at fuel fitter. Refer to paragraph 3-12. If no air, do step 3.

Step 3. Check for tow coolant temperature.

 a. If coolant temperature not low, do step 4.

 b. Replace defective thermostat. Refer to paragraph 3-6. Check

Step 4. for fuel injector nozzle dirty, defective or leaking.

 c. Clean and test fuel injector. Refer to paragraph 3-10. If not defective, do step 5.

 b. Replace fuel injector nozzle. Refer to paragraph 3-10. Check

step 5. for improper fuel injection timing.

 c. Check fuel injection timing. Refer to paragraph 3-13.1. If fuel injection timing is correct, do step 6.

 b. Adjust fuel injection timing. Refer to paragraph 3-13.2.

Step 6. Check for defective fuel injection pump.

 c. Remove and test fuel injection pump. Refer to paragraph 3-11.5. If not defective, do step 7.

 b. Replace fuel injection pump. Refer to paragraph 3-11. Check

step 7. for valves improperly adjusted.

 c. Adjust valves. Refer to paragraph 3-26.6. if properly adjusted, do step 8.

Table 3-2. Troubleshooting - Continued

MALFUNCTION
    TEST OR INSPECTION
        CORRECTIVE ACTION

    Step 8.   Check for low engine compression.

              a. Perform engine compression check. Refer to paragraph 3-27.7.

              b. If engine defective, repair or replace engine.

---

6.   **ENGINE DOES NOT DEVELOP FULL POWER.**

    Step 1.   Check for blocked air intake system.

            Remove blockage as found. If no blockage is found, do step 2.

    Step 2.   Check for air in fuel system lines.

    Step 3.   Engine overheated.

            Refer to Engine Overheating, Malfunction 7. If not a problem, refer to Malfunction 5.

---

7.   **ENGINE OVERHEATING.**

    Step 1.   Inspect coolant level.

              a. Check engine for coolant leaks. If engine has no leaks, do step 2.

              b. Repair coolant leaks.

    Step 2.   Check for defective thermostat.

               a. If thermostat is suspected of being defective, replace thermostat. Refer to paragraph 3-6.

              b. If engine continues to overheat, do step 3. Check

    Step 3.   for defective water pump.

              c. Remove and check water pump for damage. If not defective, do step 4.

              b. Replace defective water pump. Refer to paragraph 3-7. Check

    Step 4.   improper fuel pump injection timing.

            Adjust fuel injection pump timing. Refer to paragraph 3-13.2.

---

8.   **EXCESSIVE OIL CONSUMPTION.** Step

    1.   Check for oil leakage.

              a. Inspect engine for oil leaks. If no leaks, do step 2.

              b. Repair or replace defective components. Step

    2.   Check for blocked air intake system.

            Remove blockage as found. If no blockage, do step 3.

    Step 3.   Check for defective intake or exhaust valve seals or valve guides.

              a. Repair or replace defective components. Refer to paragraph 3-27.4.

              b. Disassemble and inspect valve seals and guides. Refer to paragraph 3-27.2. If not defective, replace engine.

Table 3-2. Troubleshooting - Continued

MALFUNCTION
    TEST OR INSPECTION
        CORRECTIVE ACTION

Step 1.   Check for improper grade of oil.

    a. Refer to end item lubrication order. If proper grade of oil, do step 2.

    b. If improper grade of oil, refer to end item maintenance manual and change oil and filter.

step 2.   Check for engine running hot.

    c. Refer to Engine Overheating, Malfunction 7, in this table.

    b. If not running hot, do step 3.

step 3.   Check for defective relief valve.

    c. Test oil pressure relief valve. Refer to paragraph 3-15.3. If setting is normal, do step 4.

    b. Replace defective relief valve. Refer to paragraph 3-15. Check

Step 4.   for clogged oil pump strainer.

    Remove and clean strainer. Refer to paragraph 3-18. If not clogged, do step 5.

Step 5.   Check for defective oil pump.

    a. Remove and inspect oil pump for defective parts. Refer to paragraphs 3-18.1 and 3-18.2. If not defective, do step 6.

    b. Replace oil pump. Refer to paragraph 3-18.

Step 6.   Check for worn rocker arm bushings.

    c. Replace rocker arm bushings. Refer to paragraph 3-26.

    b. Remove and inspect rocker arm bushings. Refer to paragraphs 3-26.1 and 3-26.3. If not worn, replace engine.

Step 1.   Check for leak in fuel system.

    a. Check fuel system for leaks. If no leaks, do step 2.

    b. Repair fuel system.

Step 2.   Check for blocked air intake system.

    Remove blockage as found. If no blockage is found, do step 3.

Step 3.   Check for defective fuel injector.

    a. Remove, clean, and test fuel injectors. Refer to paragraph 3-10. If not defective, do step 4.

    b. Replace fuel injectors. Refer to paragraph 3-10. Check

Step 4.   for improper fuel injection timing.

    c. Check fuel injection pump timing. Refer to paragraph 3-13.1. If fuel injection timing is correct, do step 5.

    b. Adjust fuel injection pump timing. Refer to paragraph 3-13.2.

Table 3-2. Troubleshooting - Continued

MALFUNCTION
TEST OR INSPECTION
CORRECTIVE ACTION

Step 5.   Defective fuel injection pump.

a.Remove and test fuel injection pump. Refer to paragraph 3-11.5. If fuel injection pump not defective, go to step 6.

b.Replace fuel injection pump. Refer to paragraph 3-11. Check

Step 6.   for low engine compression.

a.Perform engine compression check. Refer to paragraph 3-27.7. If compression good, do step 7.

b.If engine defective, repair or replace engine. Check

Step 7.   for valves improperly adjusted.

Adjust valves. Refer to paragraph 3-26.6.

11.   BLACK OR GRAY SMOKE.

Step 1.   Check for blocked air intake system.

Remove blockage as found. If no blockage is found, do step 2.

Step 2.   Check for defective fuel injectors.

a.Test fuel injectors. Refer to paragraph 3-10. If not defective, do step 3. b.Re-

place fuel injectors. Refer to paragraph 3-10. Check for improper fuel injection

Step 3.   pump timing.

a.Check fuel injection pump timing. Refer to paragraph 3-13.1. If fuel injection timing is correct, do step 4.

b.Adjust fuel injection pump timing. Refer to paragraph 3-13.2. Defec-

Step 4.   tive fuel injection pump.

a.Remove and test fuel injection pump. Refer to paragraphs 3-11.5. b.Re-

pair or replace fuel injection pump. Refer to paragraph 3-11.

12.   BLUE EXHAUST SMOKE.

Step 1.   Check for blocked air intake system.

Remove blockage as found. If no blockage is found, do step 2. Step 2.

Check for defective intake or exhaust valve seals or valve guides. a.Repair or replace de-

fective components. Refer to paragraph 3-27.4.

b.Disassemble and inspect valve seals and guides. Refer to paragraph 3-27.2. If not defective, replace engine.

Table 3-2 Troubleshooting - Continued

MALFUNCTION
    TEST OR INSPECTION
        CORRECTIVE ACTION

13.   ENGINE KNOCKS.

    Step 1.   Improper fuel injection pump timing.

        a. Check fuel injection pump timig. Refer to paragraph 3-13.1. If fuel injection timing is correct, do step 2.

        b. Adjust fuel injection pump timing. Refer to paragraph 3-13.2. step 2.

    Check for defective fuel injector.

        a. Remove, clean, and test fuel injectors. Refer to paragraph 3-10. If not defective, do step 3.

        b. Replace fuel injectors. Refer to paragraph 3-10. step

    3.   Check for engine compression.

        a. Check compression. Refer to paragraph 3-27.7.

        b. If engine defective, repair or replace engine.

14.   <u>ENGINE MAKES ABNORMAL NOISE</u>

    Step 1.   Check for worn or damage water pump bearing.

        a. Check for excessive water pump shaft play at cooling fan. If not excessive, do step 2.

        b. Replace defective water pump. Refer to paragraph 3-7.

    Step 2.   Defective alternator.

        a. Remove alternator and check for damage. Refer to paragraph 3-24.1. If not defective, do step 3.

        b. Repair or replace alternator. Refer to paragraph 3-24. Check

    step 3.   for improperly adjusted valves.

    check valve clearance and adjust as necessary. Refer to paragraph 3-26.6. If properly ad-justed, do step 4.

    step 4.   Check for damaged rocker arm.

        a. Inspect rocker arm for damage. Refer to paragraph 3-26.3. If not damaged, do step 3.

        b. Repair or replace damaged rocker arm. Refer to paragraph 3-26. Check

    step 5.   for loose flywheel.

        c. Tighten bolts to specified torque. Refer to paragraph 3-31.3.

        b. Check tightness of flywheel attaching bolts. If not loose, repair or replace engine.

Table 3-2. Troubleshooting - Continued

MALFUNCTION
> **TEST OR INSPECTION**
>> **CORRECTIVE ACTION**

---

15. ENGINE MAKES A GAS LEAKING NOISE.

Step 1. Check for loose or damaged exhaust manifold.

a. Inspect for damaged and attaching hardware. IF not damage or loose, do step 2.

b. Tighten or replace exhaust manifold. Refer to paragraph 3-21.3. Step

2. Check for loose fuel injection nozzle and/or glow plugs.

a. Inspect fuel injection nozzles and glow plus for looseness. If not loose, do step 3.

b. If loose, replace washers and tighten injection nozzles (paragraph 3-10) and/or glow plugs (paragraph 3-22.4).

Step 3. Check for damage cylinder head gasket.

a. Inspect area around cylinder head gasket for evidence of gas leakage.

b. Remove cylinder head and replace gasket. Refer to paragraph 3-27.

---

16. DETONATION OR PRE-IGNITION

Step 1. Improper fuel injection pump timing.

a Adjust fuel injection pump timing. Refer to paragraph 3-13. If timing is correct, proceed to Step 2.

Step 2. Defect fuel injector nozzles.

a Test fuel injection nozzles. Refer to paragraph 3-10. If fuel injection nozzles is not defective, proceed to Step 2.

b. Replace fuel injection nozzles. Refer to paragraph 3-10. Carbon

Step 3. build-up in compression chamber.

c. Remove cylinder head and inspect for carbon build-up. Refer to paragraphs 3-27 and 3-32.

b. Remove carbon and/or replace components as necessary.

---

17. BATTERY CHARGE AMMETER SHOWS CHARGE WHEN BATTERIES ARE LOW.

Step 1. Check for broken or loose fan belt.

a Inspect fan belt. Refer to end item maintenance manual. If belt not loose or broken, do step 2.

b. Adjust or replace fan belt. Refer to end item maintenance manual. Step 2.

Test for defective battery charging alternator.

a Inspect and test battery charging alternator. Refer to paragraph 3-24.1. If alternator not de fective, do step 3.

b. Repair or replace battery charging alternator. Refer to paragraph 3-24. Step 3.

Check for breaks or loose connections in charging circuit.

if breaks or loose connections are found, repair charging circuit. Refer to end item mainte-nance manual.

Table 3-2. Troubleshooting - Continued

MALFUNCTION
    TEST OR INSPECTION
        CORRECTIVE ACTION

---

18.   BATTERY CHARGED AMMETER SHOWS EXCESSIVE CHARGING AFTER PROLONGED PERIOD.

    Step 1.   Check for defective batteries.

        aTest batteries. Refer to end item maintenance manual. If batteries not defective, do step 2.

        b. Replace batteries. Refer to and item maintenance manual. Step

    2.   Test for defective battery charging alternator.

        a. Inspect and test battery charging alternator. Refer to paragraph 3-24.3. If alternator not defective, do step 3.

        b. Repair or replace battery charging alternator. Refer to paragraph 3-24. step

    3.   Check for short in charging circuit.

        If shorted, repair charging circuit. Refer to end item maintenance manual.

SECTION III. GENERAL MAINTENANCE

3-3. GENERAL

This section provides general maintenance not found in other sections of Chapter 3.

3-3.1 General Instructions.

WARNING

Where applicable, prior to performing engine maintenance, ensure bat-teries are disconnecting. Failure to observe this warning could results in severe

NOTE

Refer to end item maintenance manual for removal of any components neces-sary to gain access to engine.

a   It is strongly recommended that bolts or nuts securing cylinder heads, covers, and doors be tightened in prop-er sequence.

b.   When assembling an engine, it is always advisable nuts, bolts, and lockwashers that have been removed from high strsss locations, in particular nuts an&or bolts from connecting rods and cylinder heads should be replaced.

c.   When assembling an engine it is always advisable to apply a small quantity of new engine lubricating oil (MIL-L-2104) to all moving parts. After any maintenance work on engine has bean completed, lubricating oil and fuel levels must be checked and all safety guards installed before operating.

d. e.   When a new fan drive belt has been installed, check belt tension after first 20 hours of operation.

f.   Wear protective overalls, and keep items of loose clothing dear of all hot and moving pans. Use protective barrier cream when necessary.

g.   Whenever possible, dean components and surrounding area before removing or disassembling. Take care to exclude all dirt and debris from fuel injection equipment while it is being serviced.

h.   Some parts are cemented with gasket compound with others being dry. Before assembly, remove all traces of old gasket and compound. Take extreme care to exclude dirt from all gasket surfaces and gasket com-pound from all tapped holes unless otherwise specified.

i.   It is recommended that all oil seals are replaced once they have been removed from their original position. Seals must be installed square in housing and all lip seals must be installed with lip facing lubricant to be re-tained. A service tool should be used to install all oil seals and care must be taken to prevent damaging new seal when it passes over shafts.

   Replace ail nuts, bolts, capscrews, and studs with damaged threads. Do not use a tap-or die to repair dam-aged threads which may impair the strength and closeness of the threads and is not recommended.

j.   Do not allow grease or oil to enter a blind threaded hole as hydraulic action present when bolt or stud is screwed in could split or stress housing.

k.   To check or re-torque a bolt or nut, item is slackened a quarter of a turn and then tightened to specified value.

l.   A steel ISO metric bolt, capscrew, or nut can be identified by the letter M either on head or one hexagon flat. The strength grade will also be marked on top or one flat.

m.   On nuts with identification marks on one face the frictional area of that surface will be reduced, therefore nut should be installed with unmarked face towards component.

n.   Service took are designed to aid disassembly and assembly procedures and their use will prevent possible unnecessary damage to components. It is recommended that service tools are always used, some operations can-not be safely carried out without aid of relevant tool.

3-4.   DISASSEMBLY AND ASSEMBLY SEQUENCE FOR OVERHAUL.

The following paragraphs provide the sequence of disassembly and assembly for complete overhaul of the en-gine. Step-by-step procedures can be found in remaining sections of Chapter 3.

3-4.1. Disassembly.

WARNING

If the engine has been operating and coolant is hot, allow engine to cool before you slowly loosen filler cap and relieve pressure from cooling system. Failure to observe this warning could result in severe personal

---

WARNING

Use care when rotating engine on engine maintenance stand. If neces-sary, use a lifting device to avoid severe personal injury.

a. Drain all coolant and engine oil, refer to end item maintenance manual. Check engine oil for metal conta-mi-nates.

b. Remove fan belt and alternator, refer to paragraph 3-24.1

c. Remove starter assembly, refer to paragraph 3-23.1.

d. Remove glow plugs, refer to paragraph 3-22.1.

e. f. Remove thermostat housing with thermostat, refer to paragraph 3-6.1.

g. Remove water pump, refer to paragraph 3-7.1.

h. Remove oil fitter and oil piping, refer to paragraphs 3-15.1 and 3-16.1.

i. j. Remove fuel filter, refer to paragraph 3-9.1.

k. Remove fuel injectors and piping, refer to paragraph 3-10.1.

l. Remove fuel injection pump, refer to paragraph 3-11.1.

Remove PCV assembly, refer to paragraph 3-20.1.

Remove intake and exhaust manifolds, refer to paragraph 3-21.1.

m. Remove rocker arm cover and rocker arm assembly, refer to paragraph 3-26.1.

n. Remove pushrods, cylinder head assembly and cylinder head gasket, refer to paragraph 3-27.1. Check for bent push rods and discard cylinder head gasket.

o. Remove crank shaft pulley and front gear cover, refer to paragraph 3-25.1.

p. Remove oil pan, refer to paragraph 3-17.1.

q. Remove flywheel and flywheel housing, refer to paragraph 3-31.1

r. Remove crankcase, refer to paragraph 3-29.1.

s. Remove oil pump, refer to paragraph 3-18.1.

t. Remove idler gear assembly, refer to paragraph 3-30.1.

u. Remove pistons, connecting rods, and crankshaft, refer to paragraph 3-28.1.

v. Remove camshaft assembly, refer to paragraph 3-30.1.

w. Disassemble cylinder block assembly, refer to paragraph 3-32.1.

x. Cap/cover all openings to prevent entry of foreign material.

3-4.2. Assembly.

a. Remove all caps/covers installed during disassembly.

b. Assemble cylinder block assembly, refer to paragraph 3-32.3.

c. Install camshaft assembly, refer to paragraph 3-30.5.

d. Install crankshaft, pistons, and connecting rods, refer to paragraph 3-28.5.

e. Install idler gear assembly, refer to paragraph 3-30.5.

f. Install oil pump, refer to paragraph 3-18.3.

g. Install crankcase, refer to paragraph 3-29.3.

h. Install flywheel housing and flywheel, refer to paragraph 3-31.3.

i. Install oil pan, refer to paragraph 3-17.3.

j. Install cylinder head assembly and Push rods, refer to paragraph 3-27-6.

k. install rocker arm assembly, refer to paragraph 3-26.5.

l. install intake and exhaust manifolds, refer to paragraph 3-21.3.

m. Install PCV assembly, refer to paragraph 3-20.3.

n. Install fuel injection pump, refer to paragraph 3-11.6.

o. install front gear cover and crankshaft pulley, refer to paragraph 3-25.3.

p. Time and adjust fuel injection pump, refer to paragraph 3-13.

q. Install fuel injectors and piping. refer to paragraph 3-10.7.

r. Install glow plugs, refer to paragraph 3-22.4.

s. Install fuel filter, refer to paragraph 3-9.4.

t. Install oil filter and piping, refer to paragraphs 3-15.4 and 3-16.2.

u. Install water pump, refer to paragraph 3-7.4.

v. Install thermostat housing with thermostat, refer to Paragraph 3-6.4.

w. Install starter assembly, refer to paragraph 3-23.6.

x. Install alternator and fan belt, refer to paragraph 3-24.5.

y. Adjust valves/rocker arm assembly, refer to paragraph 3-26.6.

z. Install rocker arm cover, refer to Paragraph 3-26.5.

aa. Fill engine with proper oil and coolant, refer to end item maintenance manual.

ab. Perform normal standard engine performance checks.

SECTION IV. COOLING SYSTEM MAINTENANCE

3-5. GENERAL.

This section provides maintenance for cooling system components. Components of cooling system not mentioned in this section can be found in the end item maintenance manual.

3-6. THEMOSTAT AND HOUSING.

```
+-------------------+
|     WARNING       |
+-------------------+
```

If the engine has been operating and coolant is hot, allow engine to cool before
you slowly loosen filler cap and relieve pressure from cooling system. Failure to observe this warning could result in severe personal injury.

3-6.1. Removal

a. Drain coolant system if not already drained. Refer to end item maintenance manual.

b. If not already done, loosen hose clamp and disconnect outlet hose from outlet pipe (2, FIGURE 3-1).

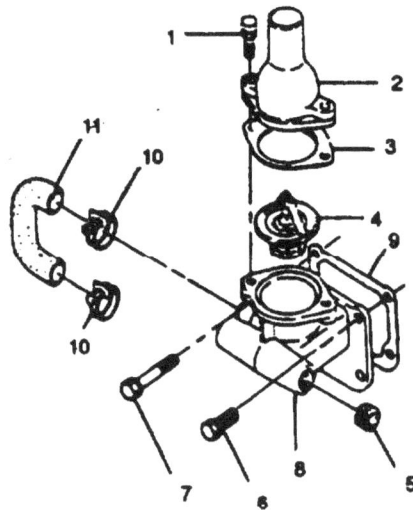

FIGURE 3-1. Thermostat and Housing

c. Remove two screws (1) securing outlet pipe (2) to housing (8); remove outlet pipe (2) and gasket (3). Discard gasket (3).

d. Lift thermostat (4) from housing.

e. Loosen two hose clamps (10) and remove bypass hose (11) from housing (8) and water pump housing.

NOTE :

Note location of two shorter screws (6) for use during installation.

f. Remove four screws (6 and 7), housing (8), and gasket (9). Discard gasket (9).

g. If necessary, remove plugs (5) from housing (8).

3-6.2. Inspection.

NOTE

If thermostat is suspected of being defective, replace thermostat.

a. Inspect thermostat for excessive wear or damage.

b. Inspect housing for cracks, corrosion, or other damage.

3-6.3. Replacement.

a. Replace thermostat if worn or damaged.

b. Replace housing if badly corroded or damaged.

3-6.4. Installation.

a. If removed, install plugs (5, FIGURE 3-1) in housing (8).

b. Apply sealing compound (FORMAGASKET2) and position housing (8) with new gasket (9) on engine block and secure with four screws (6 and 7). Install two shorter screws (6) in locations noted during removal. Tighten all four screws to 168 in-lbs (19 Nm)

c. Position thermostat (4) in housing (8).

d. Apply sealing compound (FORMAGASKET2) and position outlet pipe (2) and new gasket (3) on housing (8) and secure with two screws (1). Tighten screws to 168 in-lbs (19 Nm).

e. Install bypass hose (11) on housing (8) and water pump housing and secure with two hose clamps (10).

f. Connect outlet hose to outlet pipe (2) and secure with hose clamp.

g. Service coolant system, refer to end item maintenance manual.

3-7. WATER PUMP.

3-7.1. Removal.

┌─────────────┐
│   WARNING   │
└─────────────┘

Do not drain coolant until coolant temperature is below operating tem-pera-ture prior to removal. Severe personal injury can occur.

a. Drain coolant system if not already drained, refer to end item maintenance manual.

b. Loosen hose clamp and remove inlet hose from water pump outlet.

c. Remove fan drive belt, refer to end item maintenance manual.

d. Remove four screws (1, FIGURE 3-2) securing fan (2) spacer (3), and fan pulley (4) to water pump (7); remove fan, spacer, and pulley.

NOTE

Note location of two longer screws (5) for use during installation.

e. Remove six screws (5 and 6) securing water pump (7); remove alternator bracket (8), water pump (7) and gasket (9). Discard gasket.

3-7.2. Inspection.

a. Inspect pump rotation for abnormal noise, binding, and other abnormal conditions.

b. Inspect pump housing for cracks, corrosion, or any other damage.

3-7.3. Replacement. Replace pump assembly if inspection reveals any abnormal condition, such as bearing failure, ex-cessive end play, or abnormal rotation.

3-7.4. Installation.

```
CAUTION
```

Ensure longer bolts (5, FIGURE 3-2) are installed in the location noted during removal. Failure to observe this caution will result in pump not being secured properly.

a. Apply sealant and position new gasket (9). water pump (7) and alternator bracket (8) on engine block and secure with six screws (5 and 8). Tighten screws to 30 ft-lbs (40.6 Nm)

b. Position fan pulley (4) spacer (3) and fan (2) on water pump (7) hub and secure with four screws (1). Tighten screw to 72 in-lbs (8.0 Nm).

c. Install fan drive belt, refer to end item maintenance manual.

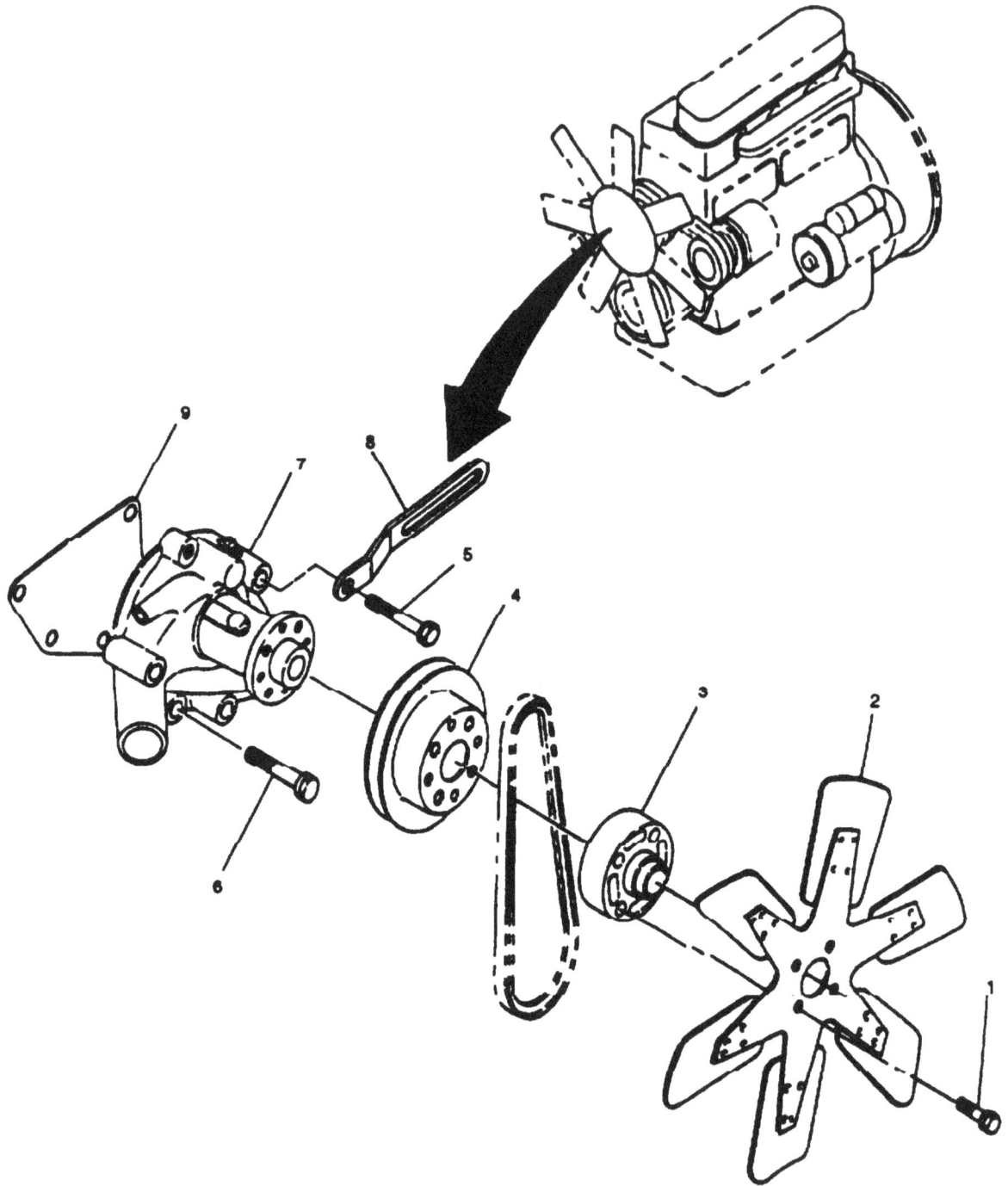

FIGURE 3-2. Water Pump

SECTION V. FUEL SYSTEM MAINTENANCE

3-8. GENERAL.

This section provides maintenance for fuel system components. Components of the fuel system not mentioned in this section can be found in end item maintenance manual.

3-9.   FUEL FILTER ASSEMBLY.

NOTE

The fuel filter assembly contains no filter element. Another fuel filter is located on the end item. Disassembly of this filter is only required to re-place preformed packings or seals if suspected of leaking.

3-9.1. Removal.

```
┌──────────────┐
│   WARNING    │
└──────────────┘
```

Diesel fuel is flammable and toxic to eyes, skin, and respiratory tract. Skin/eye protection required. Avoid repeated/prolonged contact. Good general ventilation is normally adequate.

NOTE

Drain fuel into suitable container.

a. Loosen drain plug (1, FIGURE 3-3) on bottom of filter and drain fuel.

b. Remove fluid passage bolt (2) securing fuel tube (4) to top centerfitting of filter body (15). Remove and discard two seals (3).

c. If necessary to remove fuel tube (4) remove fluid passage bolt (5) securing tube to fuel injector pump. Remove and discard two seals (6). Cap openings.

d. Disconnect fuel line from fitting (25) on side of filter body (15). Cap openings.

e. Remove fluid passage bolt (7) securing fuel tube (9) to fluid passage bolt (7). Remove and discard two seals (8). Cap openings.

f.   Disconnect fuel return line (10) to facilitate removal of fluid passage bolt (11). Cap openings.

g. Remove fluid passage bolt (11) securing fuel tube (13) to fitting on top of filter body (15). Removeand discard two seals (12).

h. Remove two bolts (14) securing filter body (15) to bracket (19); remove filter assembly.

i.   If necessary, remove two bolts (16), washers (18) and lockwashers (17) securing bracket (19); remove bracket. Discard lo&washers (17).

3-9.2. Disassembly.

a. Remove drain plug (1, FIGURE 3-3) and preformed packing (20) from thru-bolt (21). Discard preformed packing (20).

b. Remove thru-bolt (21) and remove preformed packing (22), filter bowl (23), and preformed packing (24) from body (15). Discard preformed packings (22 and 24).

c. If replacing entire filter assembly, remove and retain fitting (25).

3-9.3. Assembly.

a. Position new preformed packing (22, FIGURE 3-3) on thru-bolt (21). Position new preformed packing (24) and filter bowl (23) on body (15) and secure with thru-bolt (21). Tighten thru-bolt to 22 ft-lbs (29.8 Nm).

b. Install new preformed packing (20) and drain plug (1) into thru-bolt (21).

c. If removed, install fitting (25).

3-9.4. Installation.

a. If removed, position bracket (19, FIGURE 3-3) on engine block and secure with two bolts (16). new lockwashers (17), and washers (18).

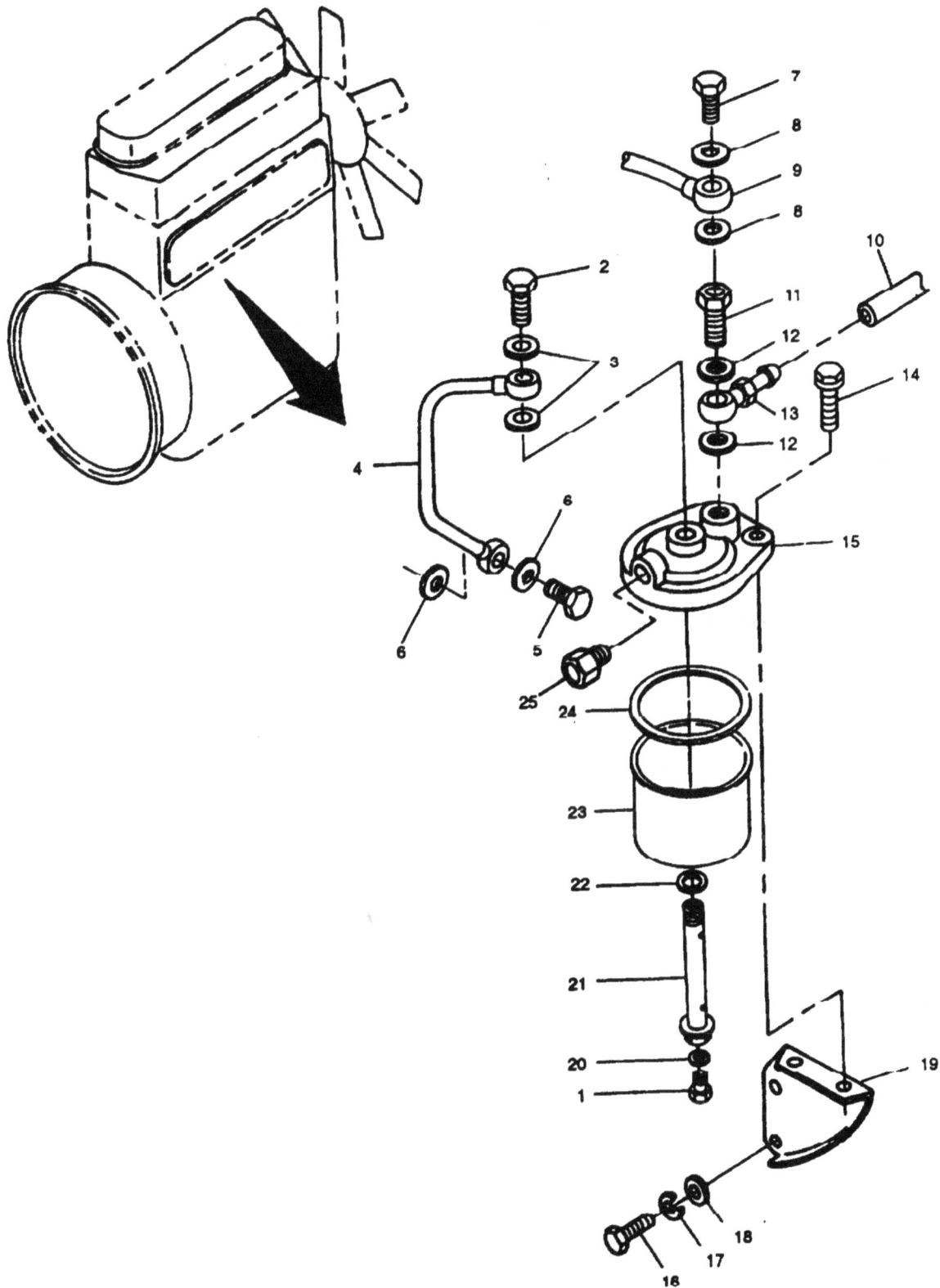

FIGURE 3-3. Fuel Filter Assembly

b. Position filter assembly on mounting bracket and secure filter body (15) with two bolts (14). Do not tighten bolts until fuel tubes have been connected to filter and tightened.

c. Remove caps and install fuel return line (10).

d. Attach fuel tube (13) to fitting on top of filter body (15) with two new seals (12) and fluid passage bolt (11). Tighten fluid passage bolt (11) to 132 in-lbs (14.9 Nm).

e. Remove caps and attach fuel tube (9) to fluid passage bolt (11) with two new seals (8) and fluid passage bolt (7).

f. Remove caps and connect fuel line to fitting (25) on side of filter body (15).

g. If removed from fuel injection pump, remove caps and attach one end of fuel tube (4) to pump with two new seals (6) and fluid passage bolt (5).

h. Attach other end of fuel tube (4) to top center fitting of filter body (15) with two new seals (3) and fluid passage bolt (2).

i. Tighten fluid passage bolts (2, 5, 7, and 11) to 132 in-lbs (14.9 Nm).

j. Tighten filter mounting bolts (14).

k. Ensure filter drain plug (1) is tight.

l. Bleed engine fuel system, refer to paragraph 3-12.

Operate engine and check for leaks.

3-10. FUEL INJECTORS AND PIPING. 3-10.1.

WARNING

Fuel is flammable. Keep fuel away from heat and open flame. Failure to

CAUTION

Do not apply excessive force to injection fuel tubes. Failure to observe this caution could damage the equipment.

NOTE

Drain fuel into suitable container.

a. Loosen sleeve nut on fuel tubes (1, 2, 3, and 4, FIGURE 3-4) at fuel injection pump

b. Loosen sleeve nut on fuel tubes (1, 2, 3, and 4) at fuel injector.

c. Remove three clamp assemblies (5) and remove fuel tubes (1, 2, 3, and 4).

d. Cap all openings to prevent entry of foreign material.

e. At each fuel injector, remove fluid passage bolt (8) and two seals (7) securing fuel tube (10). At fuel filter, remove fluid passage bolt (8) and two seals (9); remove fuel tube (10). Discard seals (7 and 9). Cap openings.

CAUTION

Handle each injector carefully to prevent damage to nozzle tip.

f. Unscrew each fuel injector (11) from block and remove washer (12) and washer (13) from each injector port. Discard washers (12 and 13). Cap or plug all openings.

3-10.2. Disassembly (Fuel Injector).

a. Loosen and remove cap nut (14, FIGURE 3-4) and gasket (15). Discard gasket (15).

b. Remove adjusting screw (16).

c. Remove washer (17), spring (18), and pushrod (19) from holder (25).

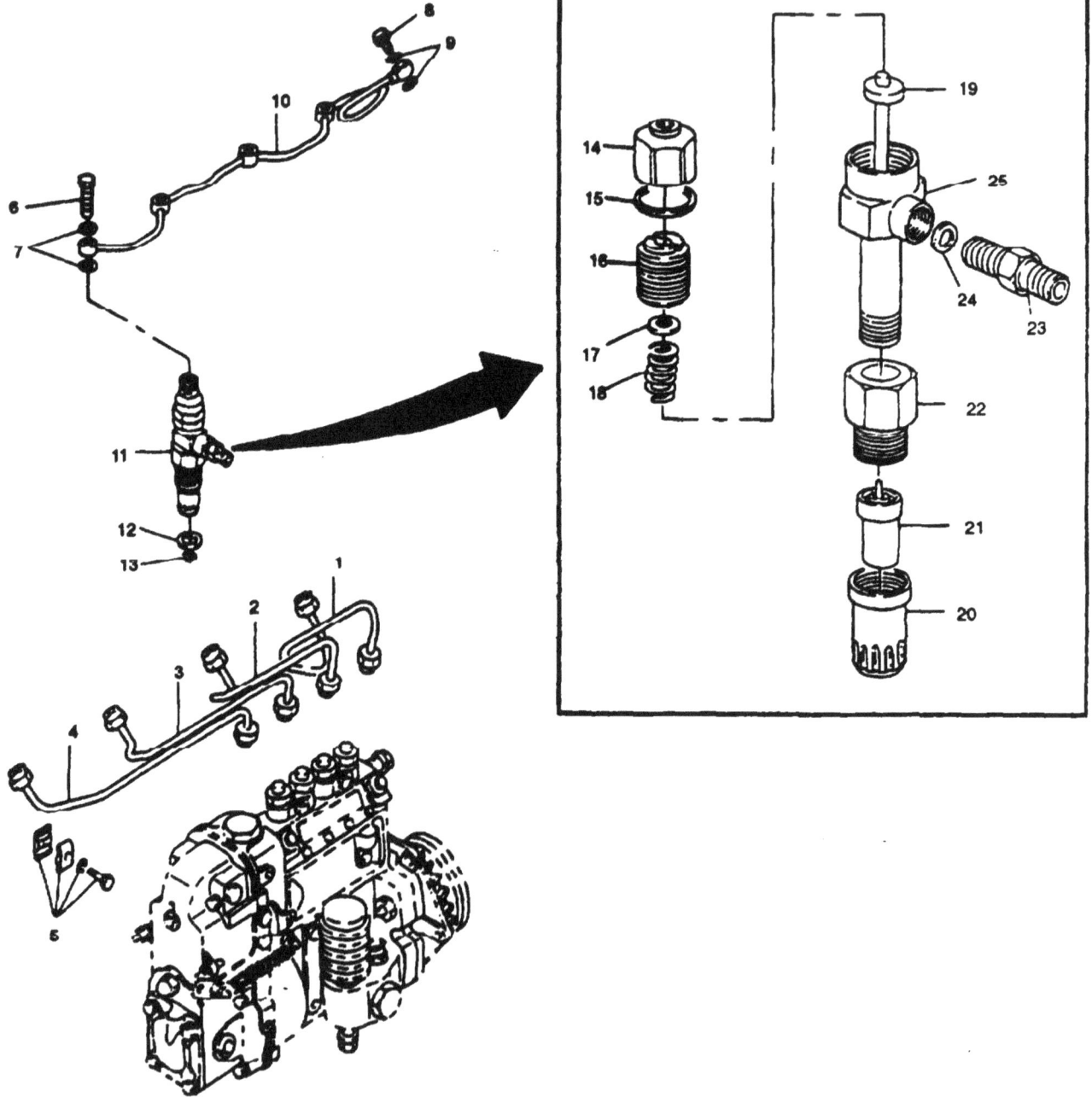

FIGURE 3-4. Fuel Injectors and Piping

d. Remove retainer nut (20) and nozzle (21) from holder (25).

e. If necessary, remove screw (22), connector (23), and gasket (24) from holder (25). Discard gasket (24).

3-10.3. Inspection.

3-10.3.1 Fuel Injector.

```
┌─────────────────┐
│     CAUTION     │
└─────────────────┘
```

The nozzle needle valve and nozzle body combinations must not be in-terchanged otherwise, injectors will not operate property.

a. Remove nozzle needle from nozzle body.

b. Clean fuel injector, refer to paragraph 3-10.4.

c. Insert nozzle needle in nozzle body and check for smooth movement of needle inside the body. If needle does not move smoothly after cleaning, replace nozzle.

d. Check nozzle body and nozzle needle for damage or deformation. If either is discovered, replace nozzle as-sembly.

e. Check both pushrod seating positions (nozzle needle and spring seat) for excessive wear and cracking. If either is dis-covered, replace pushrod.

f. Inspect threaded components for thread damage and replace as necessary.

g. Inspect spring for cracks or missing coil.

3-10.3.2. Piping.

a. Inspect all fuel tubes (piping) for wear, kinks, or fitting damage. Replace as necessary.

b. Inspect fluid passage bolts for clogged passages. Clean or replace as necessary.

3-10.4. Cleaning.

```
WARNING
```

Cleaning solvent is flammable and toxic to eyes, skin, and respiratory tract. Skin/eye protection required. Avoid repeated/prolonged contact. Good general ventilation is nom-rally adequate.

a. Soak all parts except nozzle in a dry cleaning solvent (P-D-680) and wipe off all excess residue with a soft cloth. Use wire brush to clean excessively dirty parts.

```
CAUTION
```

Clean nozzle holder mating surfaces with a piece of hard wood or a soft cloth otherwise, damage will occur. Do not use a metal brush since scars could cause leakage.

b. Remove nozzle needle valve from nozzle body. Holding the needle valve by stem, only clean shaft section and seat surface of needle valve with a piece of hardwood dipped in engine lubricating oil (ML-L-2104) or a clean soft cloth.

c. If a large amount of carbon residue remains and nozzle can not be cleaned well enough, dip nozzle in carbon cleaner and repeat step b.

```
WARNING
```

Compressed air used for cleaning can create airborne particles that may enter the eyes. Pressure will not exceed 30 psig (207 kPa). Eye protec-

```
CAUTION
```

tion required.

d. Wash all parts in dean diesel fuel. Dry with compressed air. 3-10.5.

Assembly (Fuel Injector).

a. If removed, install screw (22, FIGURE 3-4), new gasket (24), and connector (23) in holder (25).

b. Install nozzle (21) and retainer nut (20) on holder (25).

c. Install pushrod (19), spring (18), and washer (17) in holder (25).

d. Install adjusting screw (16).

NOTE

New gasket (15) and cap nut (14) will be installed after adjustment, paragraph 3-10.6.

3-10.6. Test/Adjustment.

NOTE

Perform step a through c if cap nut (14, FIGURE 3-4) is not loose.

a. Place fuel injector in a vise and clamp on holder (25).

b. Remove cap nut (14).

c. Remove fuel injector from vise.

d. Attach fuel injector to a fuel injector tester. Point fuel injector into a clear container.

WARNING

Keep body clear of test spray. Fluid can be injected into bloodstream

e. Open valve on fuel injector tester and operate lever at one stoke per second.

f. Spray should start at 1706 psi (120 kg/cm^2) and a well atomized spray pattern (refer to FIGURE 3-5) should be visible.

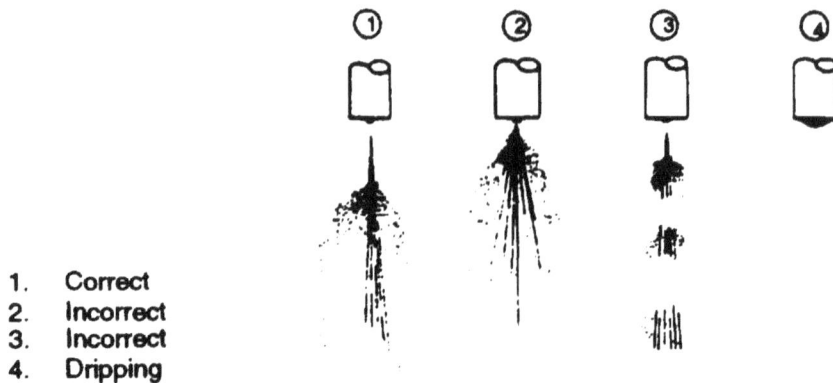

1. Correct
2. Incorrect
3. Incorrect
4. Dripping

FIGURE 3-5. Fuel Injector Spray Pattern

g. Turn adjusting screw (16, FIGURE 3-4) as necessary to obtain starting pressure and spray pattern in step f.

h. Repeat step e and operate level to hold pressure at 290 psi (20.4 kg/cm^2) below opening pressure.

i. No drops should fall from fuel injector within 10 seconds.

j. Remove fuel injector from tester and place in a vise.

k. Install capnut (14) and tighten to 25.5 ft-lbs (34.0 Nm).

l. Remove fuel injector from vise.

3-10.7. Installation.

NOTE

If existing fuel injectors are being used, ensure they are installed in loca-tions as marked during removal.

a. Remove caps/plugs and install new washers (12 and 13, FIGURE 3-4) in each injector port.

b. Install each fuel injector (11) and temporarity tighten them. Final tightening will be done after fuel tubes are connected.

Remove caps and position fuel tubes (1, 2, 3, and 4) between fuel injection pump and their respective fuel injector (11). Tighten sleeve nuts on each end evenly a little at a time, to reduce undue stress on the tubes, to 27 ft-lbs (36.6 Nm).

| CAUTION |

An improperly positioned clamp assembly (5) will result in objectionable fuel pulsing noise and possible fuel tube rupture.

d. Carefully position and tighten each of the three clamp assemblies (5).

e. f. Tighten fuel injector screw (22) to 54 ft-lbs (73.2 Nm).

g. Remove caps and position fuel tube (10) on each fuel injector (11) and at fuel filter. Secure to fuel fitter with fluid passage bolt (8) and two new seals (9). Secure to each fuel injector (11) with fluid passage bolt (6) and two new seals (7). Tighten each fluid passage bolt to 132 in-lbs (14.9 Nm)

h. Bleed engine fuel system, refer to paragraph 3-12.

Operate engine and check for leaks and proper operation.

3-11. FUEL INJECTION PUMP.

3-11.1. Removal.

| WARNING |

Fuel is flammable. Keep fuel away from heat and open flame. Failure to

NOTE

Drain fuel into suitable container.

b. Disconnect fuel tubes (1, 2, 3, and 4, FIGURE 3-4) from delivery valve ports on fuel injection pump (3, FIGURE 3-7). Cap all openings.

c. Disconnect fuel tubes from top and bottom of fuel feed pump (10). Cap openings.

d. Remove fluid passage bolt (1, FIGURE 3-8) from top front of fuel injection pump securing fuel tube from fuel fitter. Remove and discard two seals (2). Cap openings.

e. Remove fluid passage bolt (3) from side of fuel injection pump securing oil feed tube. Remove and discard two preformed packings (4).

f. Remove/disconnect engine speed control system from fuel injection pump. Refer to end item maintenance manual.

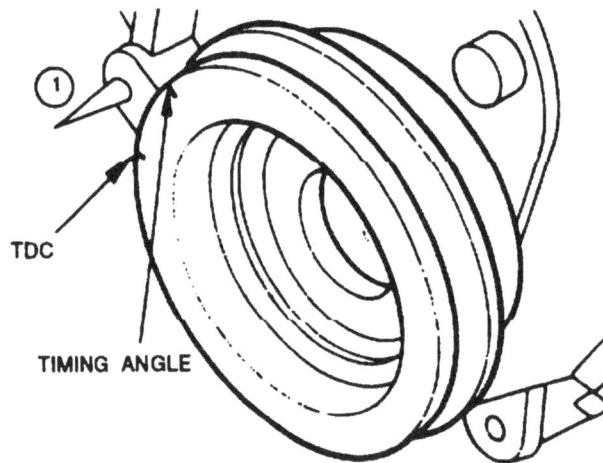

TDC

TIMING ANGLE

g. Rotate crankshaft pulley to align Top Dead Center (TDC) mark with pointer, refer to FIGURE 3-6.
FIGURE 3-6. Iop Dead Center

h. Remove four screws (1, FIGURE 3-50) securing cover (2) to front gear cover (9); remove cover (2) and pre-

i. Verify alignment of Z marks on timing gears.

NOTE

Note location of two longer bolts (2, FIGURE 3-7) for use during installa-tion.

j. Support weight of fuel injection pump (3) and remove four bolts (1) and two bolts (2) securing pump to engine front plate. Remove pump (3) and packing (4). Discard packing (4).

3-11.2. Disassembly.
Preparation.

a. Keep work area clean and orderty.

b. Record performance data for later reference. This data will facilitate detection and diagnosis of malfunctions and defects, if any.

c. Lay out parts in order of disassembly. This will aid in reassembling pump.

d. Clean outer surface of pump. Remove all grease and dirt.

e. Drain lubricating oil from pump.

3-11.2.2. Procedure.

a. Mount fuel injection pump in soft-jawed vise.

b. Attach spanner (157916-5320) to timer assembly holder (7, FIGURE 3-7), remove nut (5), and lo&washer (6). Discard lo&washer (6).

c. Attach extractor (157926-6220) and spanner (157916-5320) to timer assembly (7) and remove timer assembly.

d. Remove four nuts (5, FIGURE 3-8), lockwashers (6), and washers (7) securing bracket (8) to pump; remove bracket. Remove and discard preformed packing (10).

e. Remove pump from vise. Mount bracket (157944-7820) on universal vise (157944-8520).

f. Securely attach fuel injection pump to bracket using four bolts.

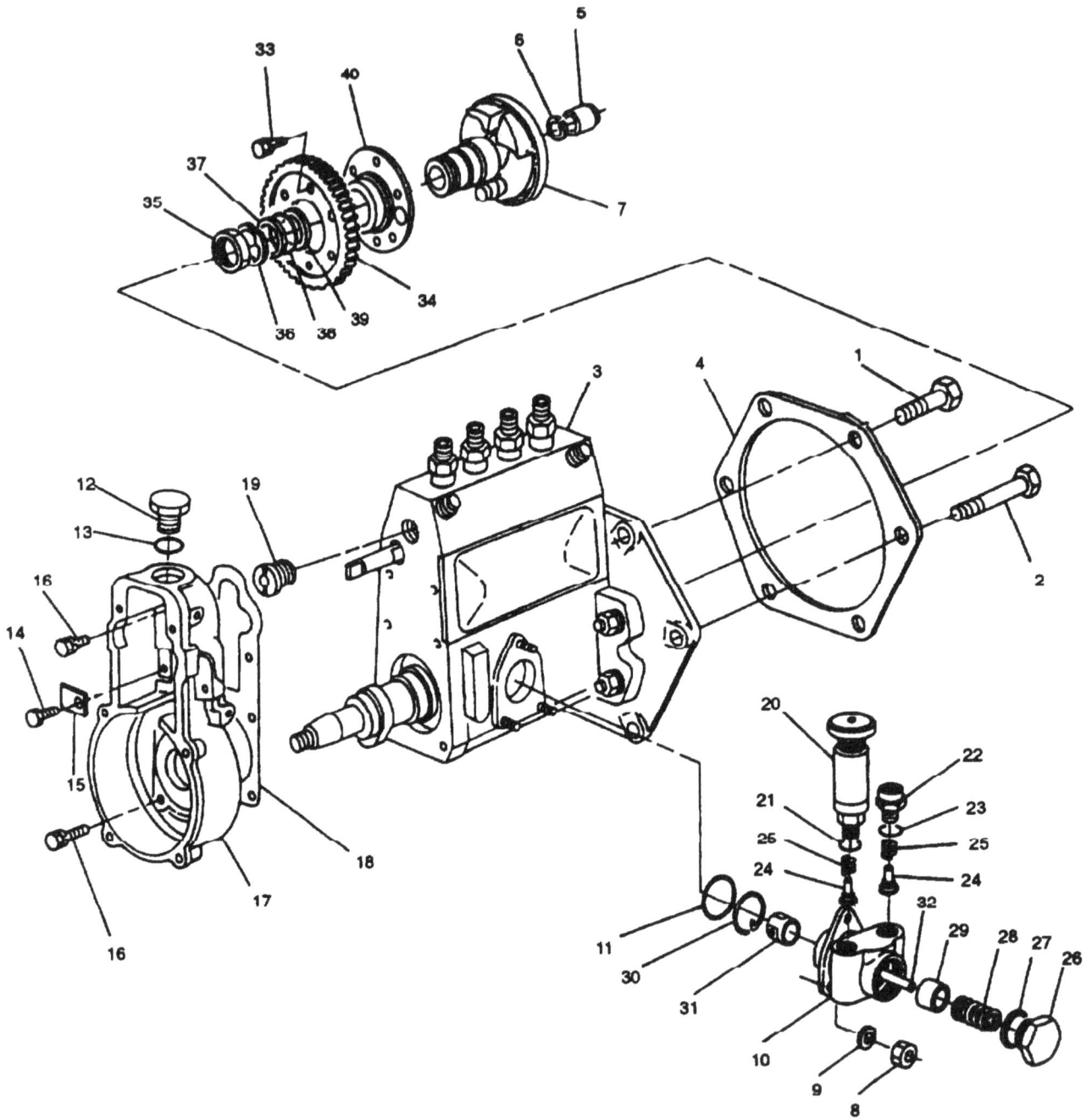

FIGURE 3-7. Fuel Feed Pump and Timer Assembly

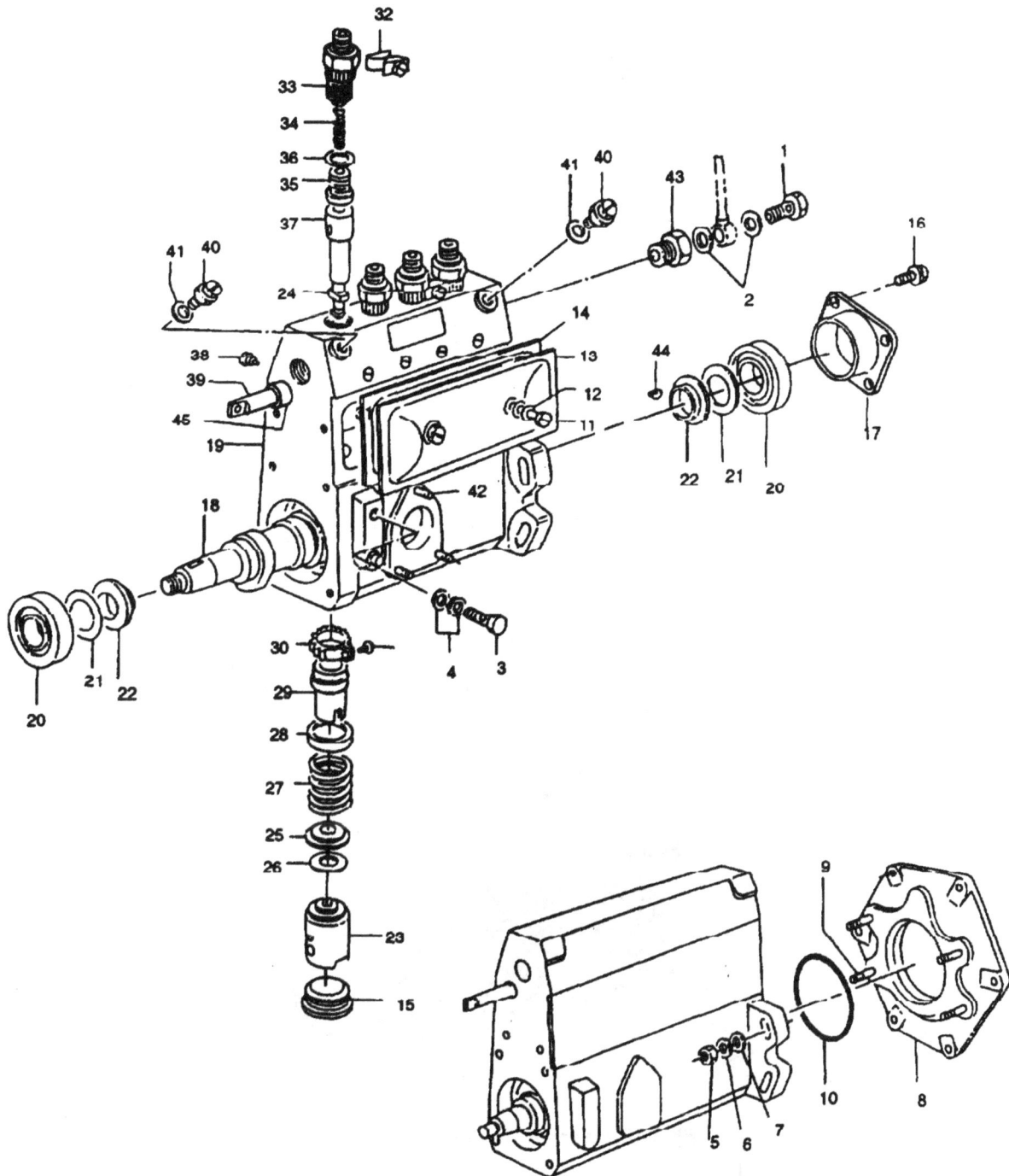

FIGURE 3-8. Fuel Injection Pump Assembly

g. Attach coupling (157842-4420) to camshaft and hold coupling using spanner (157916-5420). Tighten coupling nut.

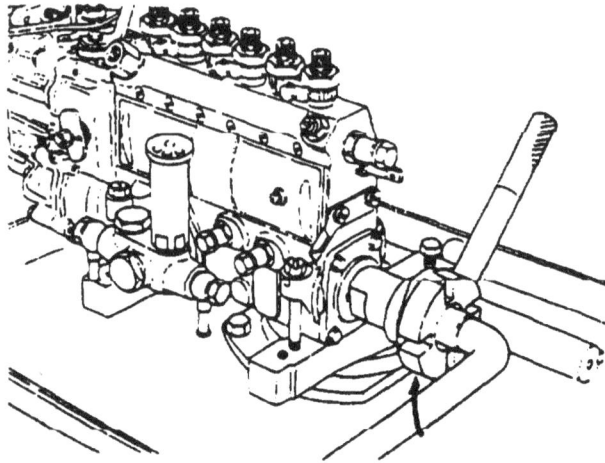

FIGURE 3-9. Attaching Coupling (Typical)

h. Remove three nuts (8, FIGURE 3-7) and lockwashers (9) securing feed pump (10) to injection pump; remove feed pump (10) and preformed packing (11). Discard lockwashers (9) and preformed packing (11).

i. Remove two screws (11, FIGURE 3-8) securing cover (12) to side of pump; remove cover. Remove and discard two gaskets (13) and gasket (14).

j. Rotate camshaft (18) until tappet (23) is at TDC.

k. Insert tappet holder (fabricate four each in accordance with FIGURE D-1, Appendix D) into small hole on each tappet (23) to disconnect tappet (23) from camshaft (18). Continue to rotate camshaft until one tappet holder is installed in each of four tappets, refer to FIGURE 3-10.

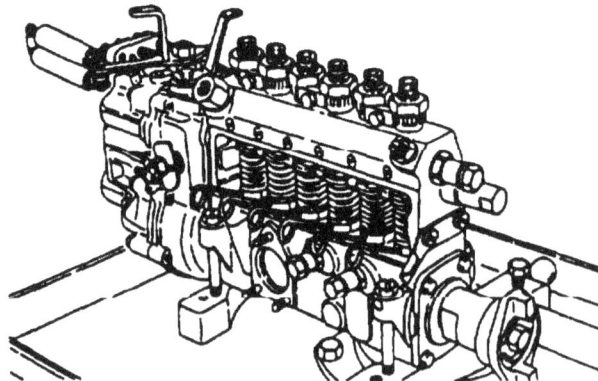

FIGURE 3-10. Inserting Tappet Holder (Typical)

l. Refer to end item maintenance manual and remove governor mechanism from governor housing (17, FIG-URE 3-7).

m. Remove plug (12) and preformed packing (13) from top of governor housing (17).

n. Remove bolt (14) with retainer (15) and six screws (16) securing housing (17) to end of pump. By tapping housing (17) lightly with a mallet, separate it from the pump. Remove and discard gasket (18).

o. Remove adapter (19) from pump housing (3).

p. Lay pump on its side and remove four plugs (15, FIGURE 3-8) from bottom of pump, refer to FIGURE 3-11.

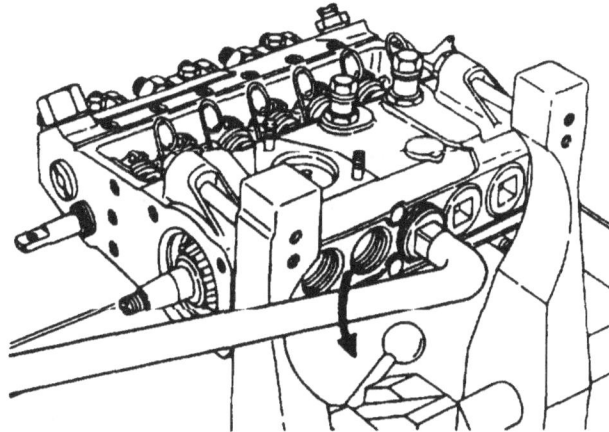

FIGURE 3-11. Removing Screw Plug (Typical)

q. Secure pump in upright position.

r. Remove coupling from camshaft using spanner wrench and socket wrench, refer to FIGURE 3-12.

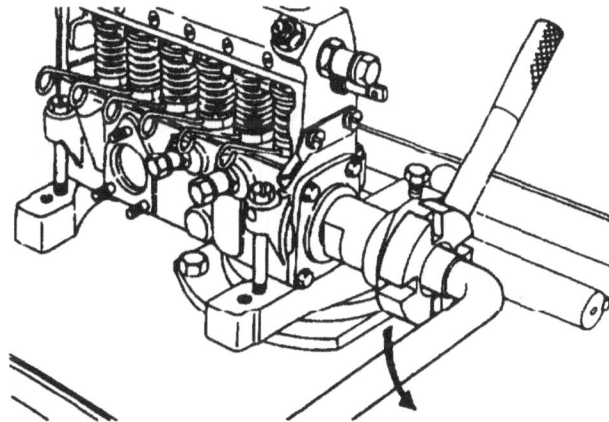

FIGURE 3-12. Removing Coupling (Typical)

NOTE

Note location of oil flow hole in cover (17, FIGURE 3-8).

s. Remove four screws (16) securing cover (17). Tap governor end of camshaft (18) with a mallet and remove cover (17) using two screwdrivers to pry cover out of housing.

t. Remove camshaft (18) from drive end of housing (19).

NOTE

Note location of rings (22).

u. If necessary, remove bearings (20), shims (21), and rings (22) from each end of camshaft.

v. Lay pump (19) on its side.

NOTE

Steps w through ac apply to each of the pumping chambers.

w. Insert tappet installer (157921-0120) into bottom plug hole, refer to FIGURE 3-13.

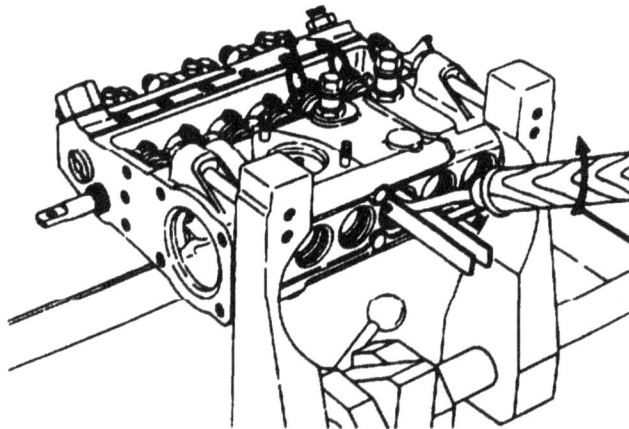

FIGURE 3-13. Removing Tappet Holder (Typical)

x. Push tappet (23, FIGURE 3-8) using the installer and remove tappet holder installed in step k.

y. Insert clamp (157931-6120) through camshaft opening, clamp tappet (23) and remove it from housing, refer to FIGURE 3-14.

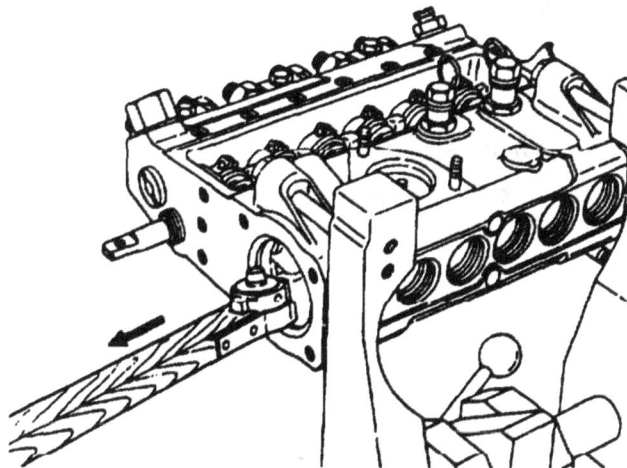

FIGURE 3-14. Removing Tappet (Typical)

CAUTION

Be careful not to let plunger (24) come in contact with other parts. Place them in clean diesel fuel in their order of removal.

z. Using mechanical finger through bottom plug hole, remove plunger (24), lower spring seat (25) and spacer ring (26)

aa. Remove spring (27) through bottom plug hole.

ab. Remove upper spring seat (28) and then remove cylinder sleeve (29), sleeve pinion (30), and screw (31) as an assembly.

NOTE

If cylinder sleeve assembly is to be disassembled, scribe a mark on sleeve (29) and sleeve pinion (30) to aid in assembly.

ac. If necessary, loosen screw (31) and remove sleeve pinion (30) from cylinder sleeve (29).

ad. Position pump in an upright position.

ae. Remove two plate assemblies (32).

af. Loosen four delivery valve bodies (33).

Remove delivery valve body (33) and spring (34).

ag. Screw extractor (1579299620) into delivery valve (35). Push down the pin and remove delivery valve (35) and gasket (36), refer to FIGURE 3-15.

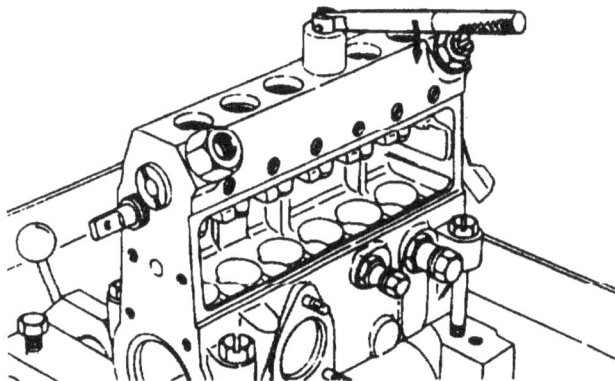

FIGURE 3-15. Removing Delivery Valve

CAUTION

Be sure to keep each plunger (24, FIGURE 3-8) with its corresponding plunger barrel (37). If plungers and barrels are not matched, clearances between them may not meet standard values.

ai. Using a finger, push plunger barrel (37) out of housing, refer to FIGURE 3-16. Attach it to its corresponding plunger removed in step z and place it in clean diesel fuel.

aj. Remove control rack guide screw (38, FIGURE 3-8). ak.

Withdraw control rack (39) from pump housing.

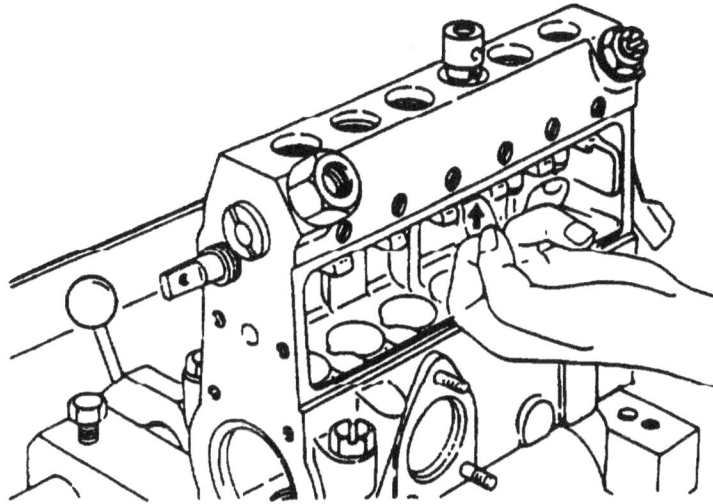

FIGURE 3-16. Removing Plunger Barrel

al. If necessary, remove and retain two bleed screws (40), two gaskets (41), three studs (42), and adapter (43).

3-11.2.3. Feed Pump.

CAUTION

Use vise jaw covers to avoid damage to feed pump housing (10, FIGURE 3-7).

a. Clamp feed pump in a vise with priming pump (20) up.

b. Unscrew and remove priming pump (20) and preformed packing (21). Discard preformed packing.

c. Unscrew and remove adapter (22) and preformed packing (23). Discard preformed packing.

d. Remove both check valves (24) and springs (25).

e. Position pump housing in vise so plug (26) faces upward.

NOTE

Tap plug (26) with a hammer to aid in loosening it.

f.   Remove plug (26).

g.   Remove gasket (27) and spring (28). Discard gasket.

h.   Remove piston (29).

i.   Turn pump housing (10) over in vise. Remove and discard retaining ring (30).

j.   Remove tappet (31) from housing and withdraw push rod (32).

3-11.2.4.   Timer Assembly.

NOTE

Disassembly of the timer will only be necessary if timer does not meet the specifications contained in paragraph 3-11.3.3. The only disassem-bly authorized is described below.

a.   Mount base of inserter tool (157924-1620) in a vise.

b.   If not already done, install guide bolt in base of inserter.

c.   Position timer assembly (7, FIGURE 3-7) over guide bolt so that pins on inserter base enter holes in timer assembly (7) flyweight holder.

d.   Remove six screws (33) securing gear (34) to flange (40); remove gear.

e.   Flatten tab of lockplate (36).

f.   Remove nut (35), lockplate (36), washer (37), shim (38), and washer (39) securing flange (40) to holder (7). Discard lockplate.

3-11.3. Inspect.

WARNING

Diesel fuel is flammable and toxic to eye-s, skin, and respiratory tract. Skin/eye protection required. Avoid repeated/prolonged contact. Good general ventilation is normally adequate.

a. Thoroughly wash all disassembled parts in clean diesel fuel.

b. Check each part carefully. If they are damaged or worn excessively, replace them.

c. Replace all gaskets and preformed packings. Do not reuse them.

3-11.3.1. Injection Pump.

a. Check that plunger helix (FIGURE 3-17) is not damaged, that color of plunger and plunger barrel (37, FIGURE 3-8) have not changed and that Plungers move smoothly in barrels. If there are any problems, replace plunger and barrel as an assembly.

FIGURE 3-17. Plunger Helix

b. Ensure that plunger (24) slides smoothly into barrel (37) under its own weight when the assembly is inclined approximately 60 degrees. Change plunger position by rotating it and check plunger movement several times as above. If plunger does not slide smoothly, replace plunger and barrel as an assembly, refer to FIGURE 3-18.

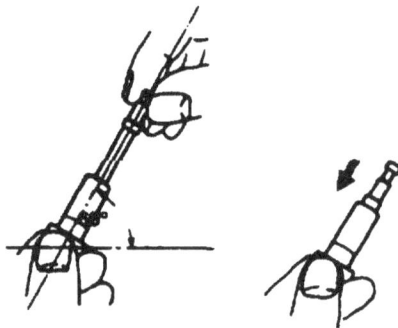

FIGURE 3-18. Checking Plunger and Barrel Movement

c. Inspect delivery valve (35, FIGURE 3-8) piston and seat for nicks, dents, and excessive wear. If any signs of damage are evident, replace entire delivery valve assembly.

d. Holding delivery valve seat, close port with finger tip. Push valve into seat and check that it is returned (by the air inside) when released, refer to FIGURE 3-19. If valve is not returned, then fit of piston in bore is too loose (due to wear) to justify reuse of this delivery valve assembly. Replace the entire delivery valve assembly.

e. Check that control rack (39, FIGURE 3-8) is not bent. If bent, replace control rack.

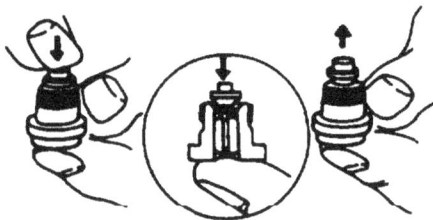

FIGURE 3-19. Delivery Valve

f. Check for excessive clearance between control rack (39) and bushing (45) that is press fitted into pump hous-ing (19). To check for end play, hold pinion and measure movement of control rack. If wear on working area of control rack and bushing is excessive and play is more than 0.012 in. (0.3 mm), replace control rack.

g. Check for excessive wear of gear teeth on control rack (39) and sleeve pinion (30). If backlash is more than 0.012 in. (0.3 mm), replace both control rack (39) and sleeve pinion (300).

h. Check that flange on plunger (24) is not damaged, refer to FIGURE 3-20. If damaged, replace plunger and barrel as an assembly.

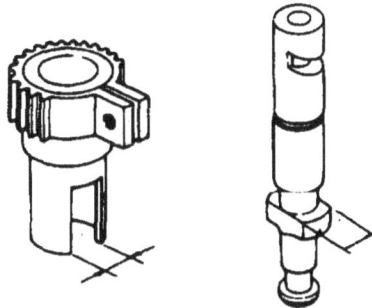

FIGURE 3-20. Checking Plunger and Barrel for Damage

i. Check cylinder sleeve (29, FIGURE 3-8) groove for damage or excessive wear. Replace if worn or damaged.

j. Check tappet (23) rollers, bushings, and pins for excessive wear, flaws, or peeling, refer to FIGURE 3-21. If damage is found, replace tappet.

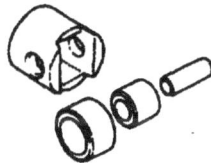

FIGURE 3-21. Tappet

k. Check top surface of lower spring seat (25, FIGURE 3-8) for damage, refer to FIGURE 3-22. If damaged, replace seat.

FIGURE 3-22. Lower Spring Seat

l.   Check camshaft (18, FIGURE 3-8) for damaged or worn cam surfaces, refer to FIGURE 3-23. Check thread damage and keyway groove on both ends. If damaged, replace camshaft.

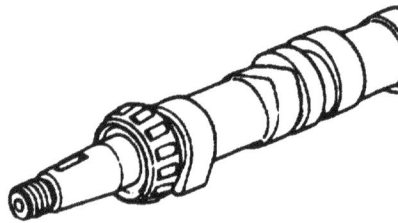

FIGURE 3-23. Camshaft

m. Check bearings (20, FIGURE 3-8) for rollers deviating from inner race or peeled, damaged, or burnt surfaces, refer to FIGURE 3-24. If found defective, refer to FIGURE 3-25 and replace defective bearings by pressing bearings from shaft and race from cover (17, FIGURE 3-8) or governor housing (17, FIGURE 3-7).

FIGURE 3-24. Camshaft Bearings

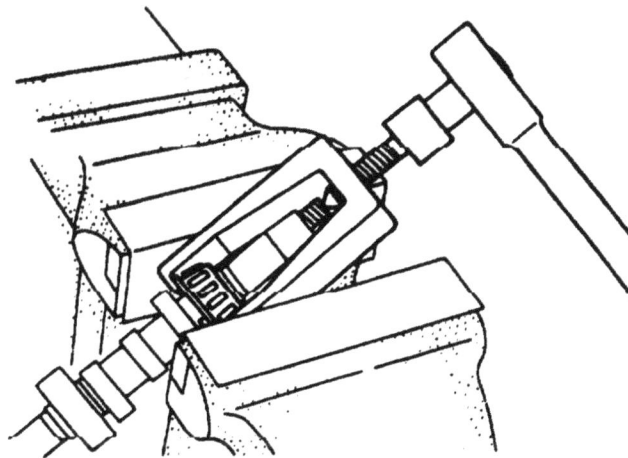

FIGURE 3-25. Bearing Removal

3-11.3.2.   Feed Pump.

    a. Inspect check valve (24, FIGURE 3-7) seats for wear or damage. If defective, replace check valve.

    b. Check piston (29) for damage or cracks. If defective, replace piston.

    c. Check tappet (31) rollers, bushing, and pin for excessive wear, flaws, or peeling. If defective, replace tappet, refer to FIGURE 3-26.

FIGURE 3-26. Feed Pump Tappet

3-11.3.3.   Tuner Assembly.

a. Visually inspect for broken springs. If springs are broken, replace timer assembly.

b. Inspect gear (34, FIGURE 3-7) for tooth wear. If defective, replace gear.

c. Using feeler gage, check clearance between washer (39) and shim (38). Clearance should be 0.0008 to 0.004 in. (0.02 to 0.1 mm). If clearance is not within tolerance, change shim thickness.

3-11.4. Assembly.
   Timer Assembly.

a. Mount base of inserter tool (157924-1620) in a vise.

b. Install holder (7, FIGURE 3-7) on base.

c. Install washer (39). shim (38), washer (37), new lockplate (36), and nut (35) on holder (7). Tighten nut 72 to 130 ft-lbs (98 to 176 Nm).

d. Using a feeler gage, ensure that clearance between washer (39) and shim (38) is 0.0008 to 0.004 in. (0.02 to 0.1 mm). If not change shim thickness.

e. Bend tab of lockplate (36) to lock nut (35).

f. Install gear (34) and screws (33). Torque screws 5.8 to 8.7 ft-lbs (7.8 to 11.7 Nm).

3-11.4.2. Feed Pump.

a. Clamp pump housing in vise, with protected jaws, so tappet opening is facing upward.

b. Insert pushrod (32, FIGURE 3-7) and tappet (31) into housing.

c. Install new retaining ring (30).

d. Turn pump housing over in vise.

e. Install piston (29) and spring (28).

f. Install new gasket (27) and plug (26). Applysealing compound (FORMAGASKET2) to threads and seating face of plug (26). Tighten plug 63 ft-lbs (86.4 Nm).

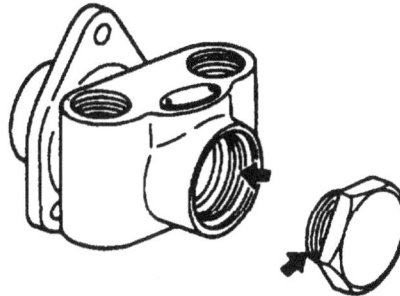

FIGURE 3-27. Feed Pump Plug

g. Position pump housing so priming pump opening faces upward.

h. Install both check valves (24) and springs (25) into pump housing.

i. Install new preformed packing (23) and adapter (22) into housing. Tighten adapter 120 in-lbs (13.6 Nm).

j. Install new preformed packing (21) and priming pump (20) into pump housing. Using 15/16 in. crows foot wrench, tighten priming pump 32 ft-lbs (43.4 Nm).

3-11.4.3.   Injection Pump.

a.   Mount pump housing (19, FIGURE 3-8) in bracket (157944-7820) mounted on universal vise (157944-8520)

b.   Install four plunger barrels (37) so that locating pins in pump housing (19) mate with grooves in barrels (37). Ensure that barrels (37) cannot manually be rotated.

NOTE

If fuel injection pump is to be tested, leave springs (24), delivery valves (35), and gasket (36) out until prestroke test is complete.

Position delivery valves (35) with new gaskets (36) on plunger barrels (37).

c.   Install four springs (34) and valve bodies (33) in pump housing. Tighten as follows:

(1) Tighten to 21.7 ft-lbs (29.4 Nm).

(2) Loosen to zero torque.

(3) Retighten to 21.7 ft-lbs (29.4 Nm).

(4) Loosen to zero torque.

(5) Finally tighten to 25.3 ft-lbs (34.3 Nm).

e.   Install control rack (39) in pump housing.

f.   Install measuring device (1057826280) on nipple (45). Use measuring device to position control rack (39) at center of travel. Install lock screw (38) in guide screw hole to lock control rack (39). Remove measuring device.

g. h.   Lay pump housing over on its side.
    If removed, position sleeve pinion (30) on cylinder sleeve (29) so that scribed marks (from disassembly) are aligned. Tighten screw (31) to 26.5 in-lbs (3.0 Nm).

Install cylinder sleeve (29) with split in sleeve pinion (30) facing upward.

i.   Install guide screw (38) into back of pump housing. Ensure control rack (39) moves smoothly.

k. Push control rack (39) fully toward drive side and then pull fully in opposite direction. Ensure that sleeve pinion (30) moves through same angle from control rack'scenter position when control rack is pushed or pulled, refer to FIGURE 3-28.

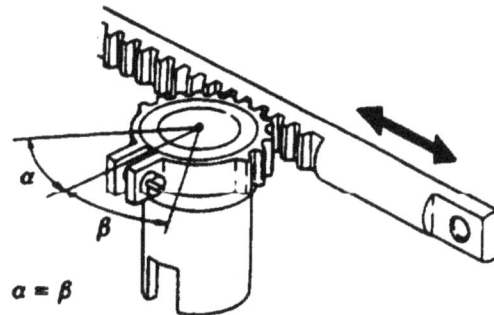

FIGURE 3-28. Checking Sleeve Pinion Movement

l. Install upper spring seat (28) and spring (27) on plunger barrel (37) through bottom plug holes.

m. Dip plunger (24) in clean diesel fuel and hold it with mechanical fingers. Insert plunger with lower spring seat (25) into barrel (37) with helix facing upward and manufacturer's mark facing downward, refer to FIGURE 3-29.

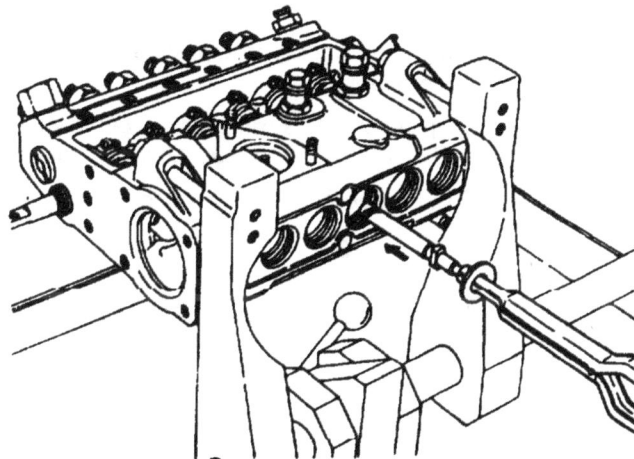

FIGURE 3-29. Inserting Plunger

NOTE

Refer to FIGURE D-1, Appendix D to manufacture tappet holders.

n. Hold tappet (23, FIGURE 3-8) with clamp (157931-6120). Align tappet guide with pump housing guide groove and insert tappet (23) and spacer ring (26) through camshaft opening, refer to FIGURE 3-30.

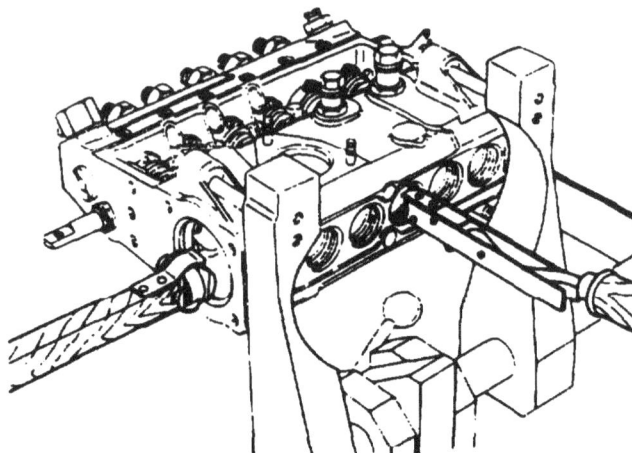

FIGURE 3-30. Inserting Tappet

o. Using tappet installer (157921-0120) through bottom plug hole, push tappet (23, FIGURE 3-8) upwards, aligning plunger flange with groove of cylinder sleeve (29).

p. When, by pushing upward on tappet (23), plunger (24) is completely inserted into plunger barrel (37) and cylinder sleeve (29); insert tappet holder (FIGURE D-1, Appendix D) into small hole on tappet (23) body.

q. If removed, install rings (22), shims (21), and bearings (20) on each end of camshaft (18).

NOTE

Code number (camshaft assembly mark) will be located on one end of camshaft indicating governor end.

r.   Install cover (17) so oil flow hole is facing as noted in disassembly on drive end and secure with four screws (16). Tighten screws 72 in-lbs (8.1 Nm).

s. t.   Apply adhesive (242) and install four plugs (15) in bottom of pump. Tighten plugs 47 ft-lbs (63 7 Nm)

u. v.   Place pump in upright position.

w.   Install adapter (19, FIGURE 3-7) in pump housing (3).

Insert camshaft (18, FIGURE 3-8) into pump housing from governor end.

x. y.   Position new gasket (18, FIGURE 3-7) and governor housing (17) on pump housing and secure with one bolt (14), retainer (15), and six screws (16). Tighten bolt (14) to 150 in-lbs (16.9 Nm) and screws (16) to 72 in-lbs (8.1 Nm).

z.   If removed, install new preformed packing (13) and plug (12) in top of governor housing (17).

Checkcamshaft end play. Using a dial indicator measure end play. End play should be 0.001 to 0.002 in. (0.03 to 0.05 mm). If end play is not within limits, remove camshaft and change thickness of shims (21, FIGURE 3-8) and recheck end play.

Check control rack (39) sliding resistance by tipping pump back and forth. Check that rack moves freely.

aa.   Using measuring device (1057826280) check the stroke of control rack (39). The stroke should be 0.83 in. (21 mm) from position where control rack is fully pushed toward governor to position where it is fully pulled toward drive end. If stroke is not 0.83 in. (21 mm) it will be necessary to change engagement of sleeve pinions (30) with control rack to obtain proper stroke.

ab.   Attach coupling (1578424420) to camshaft and hold coupling using spanner (157916-5420). Tighten nut on coupling.

ac.   Rotate camshaft (18) until tappet (23) is in TDC position.

ad.   Remove tappet holders (FIGURE D-1, Appendix D) and continue to rotate camshaft until all the tappet holders are removed. Confirm that camshaft rotates freely by manually rotating camshaft slowly.

ae. Refer to end item maintenance manual and install governor mechanism.

af. Install two plate assemblies (32) to valve bodies (33). Tighten screw 36 in-lbs (4.1 Nm). ag.

Perform fuel injection pump test, refer to paragraph 3-11.5.

ah. Position new preformed packing (11, FIGURE 3-7) and feed pump (10) on pump housing and secure with three new lockwashers (9) and nuts (8).

ai. Position cover (13, FIGURE 3-8) with new gasket (14) on side of pump housing and secure with two new gaskets (12) and screws (11).

aj. Position bracket (8) and new gasket (10) on drive end of housing and secure with four washers (7), new lockwashers (6), and nuts (5).

ak. Install timer assembly on camshaft and secure with new lockwasher (6, FIGURE 3-7) and nut (5). Attach spanner (157916-5320) to timer assembly to hold it while tightening nut (5) to 45 ft-lbs (61 Nm).

3-11.5. Testing.

a. If pump was not disassembled for cleaning and inspection, refer to paragraph 3-11.2.2 steps a thru d.

b. If not already installed, attach coupling (157842-4420) to camshaft and hold coupling using spanner (157916-5420). Tighten nut on coupling using socket wrench.

c. If not removed, remove feed pump, paragraph 3-11.2.2, step h.

d. If not removed, remove cover (13, FIGURE 3-8).

e. Remove pump from soft-jawed vice and mount pump on test stand.

NOTE

Refer to TABLE 3-1 for a listing of fuel pump test specifications.

f. Remove delivery valve bodies (33), springs (34), gaskets (36), and valves (35). Reinstall delivery valve bodies.

g. Connect fuel pipes and injection pipes. Refer to TABLE 3-2 for correct sizes.

h. Plug feed pump opening with plastic plug. Fill camshaft chamber with engine lubricating oil (MIL-L-2104) (80 cc). Fill governor chamber with engine oil, refer to PMCS chart (TABLE 3-1).

i. Remove control rack cap and attach measuring device (105782-6280).

j. Remove control rack guide screw (38) and lock control lever rack at 0.472 in. (12 mm) using locking screw.

k. Attach measuring device (105782-4020) to pump housing No. 2 cylinder. Measuring needle must rest on tappet guide pin.

l. Place No. 2 cylinder cam in bottom dead center (BDC) position. Set dial gage indicator to zero (0).

NOTE

Manually rotate camshaft by hand and ensure that indication of dial gage indicator varies with change in cam lift.

FIGURE 3-31. Measuring Device Installed (Typical)

m. Supply test oil to injection pump at as low a pressure as possible (2.8 psi (19.6 kPa)

TABLE 3-1. Fuel Pump Test Specifications

Injection Pump:	PES4A 101043-9170	Time Device:	EP/SCD 105621-0370

1. Test Conditions.

   Pump rotation: Counterclockwise, viewed from drive end.
   Nozzle: 105780-000 (Bosch Type No. DN12SD12T)
   Nozzle holder: 105780-2080 (Bosch Type No. EF5811/9A)
   Nozzle opening pressure: 2489 psi (175 kg/cm^2)
   Transfer pump pressure: 23 psi (1.6 Kg/cm^2)
   Injection pipe: ID 0.079 in. (2 mm) x OD 0.236 in. (6 mm) x
                      23.6 in. (600 mm) long.
   Test oil: ISO4113 or SAE Standard Test Oil (SAE J967d)
   Test oil temperature: 104 to 113°F (40 to 45°C)

2. Injection Timing.

   No. 1 plunger pre-stroke: 0.089 ± 0.002 in. (2.25 ± 0.05 mm)
   Injection order: 1-3-4-2
   Angle interval: 89.5 to 90.5 degrees
   Tappet clearance: More than 0.012 in. (0.3mm) with all cylinders.

3. Injection Quantity.

Rod Position inch (mm)	Pump Speed rpm	Injection Qty cc/1000 strokes	Max. Variation between cyl. (%)
0.472 (12)	1450	42-44	± 2.5
0.299 (7.6)	350	6.9-9.1	± 14

n. Rotate camshaft counterclockwise (as viewed from drive end) until test oil stops flowing from No. 2 injector test stand overflow pipe.

o. Measure lift (pre-stroke) of No. 2 plunger.

p. If pre-stroke is not 0.087 to 0.091 in. (2.20 to 2.30 mm), adjust as follows:

    (1) Rotate camshaft until No. 2 cam reaches TDC position.

    (2) Insert plunger spring holder (157931-4100) between lower spring seat (25. FIGURE 3-8) and tappet (23).

    (3) Remove spacer ring (26) and install spacer ring that will provide desired pre-stroke. Spacer rings are available in 0.002 in. (0.05 mm) steps.

    (4) Remeasure No. 2 cylinder's pre-stroke.

q. Set pump test stand flywheel pointer to an arbitrary angle graduation, with No. 2 cam placed in beginning of injection position.

r. Ensure test oil stops flowing from measuring device pipe in this position.

```
┌─────────────────┐
│    CAUTION       │
└─────────────────┘
```

Do not move flywheel pointer after it has been placed in this position. Re-

s. t. move measuring device from No. 2 cylinder.

Supply test oil at high pressure to injection pump.

u. v. w. Open nozzle holders overflow valves so test oil flows.

x. Measure beginning-of-injection positions in order of injection.

Rotate camshaft until test oil stops flowing from overflow pipe.

Read pointer indication on flywheel.

y. If angle interval is not between 89.5 and 90.5 degrees between each cylinder in firing order (1-3-4-2), adjust using spacer ring (26). If oil stops flowing prematurely, decrease thickness of shim, if oil stops flowing too late,

z. Confirm that camshaft rotates smoothly by manually rotating camshaft slowly aa. Install valves (35), gaskets (36), springs (34), and bodies (33), refer to paragraph 3-11.4.3.

NOTE

Ensure fuel injection pump is bled of all air before doing following steps. ab.

Set test stand for an inlet pressure of 23 psi (1.6 Kg/cm^2).

ac. Ensure control rack is at 0.472 in. (12 mm) position, pump speed at 1450 rpm and plunger strokes for 1000.

ad. Operate test stand and measure fuel injection quantity for each cylinder. Each cylinder should measure 42 to 44 cc with $\pm$ 1.08 cc between cylinders.

ae. If fuel injection quantity is not as specified, adjust as follows:

NOTE

Scribe a mark to identify starting point.

(1) Loosen sleeve pinion screw (31).

(2) Rotate cylinder sleeve (29) clockwise or counterclockwise using an appropriate tool. Rotating cylinder sleeve towards governor increases fuel injection quantity.

af. Repeat steps ab thru ae for the following settings.

(1) Pump speed 350 rpm.

(2) Plunger strokes 1000.

(3) Control rack at 0.299 in. (7.6 mm) position.

(4) Each cylinder should measure 6.9 to 9.1 cc with + 1.12 cc between cylinders. ag. Remove

rack lock screw from pump housing and reinstall control rack guide screw (38).

ah. Perform feed pump suction test as follows:

    (1) Install feed pump, paragraph 3-11.4.2.

    (2) Attach a 6.5 ft (2 m) length of 0.315 in. (8 mm) ID pipe to suction port of feed pump and insert other end into fuel container. Refer to FIGURE 3-32.

    (3) Attach pipe to discharge port of feed pump and insert other end into empty container.

NOTE

Ensure a suction head of approximately 3.3 ft (1 m).

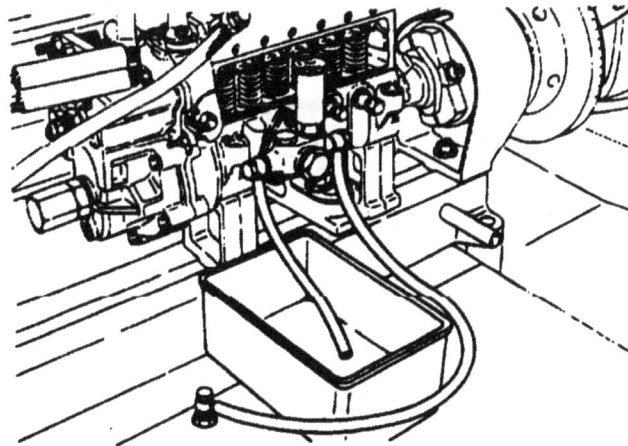

FIGURE 3-32. Feed Pump Test

    (4) Operate injection pump at 100 rpm.

    (5) Fuel oil must be discharged in less than 60 seconds. ai.

Perform feed pump priming pump test as follows:

    (1) Pump should be piped as in step ah, refer to FIGURE 3-33.

    (2) Fully loosen priming pump's screw cap.

    (3) Operate priming pump at 60 to 100 strokes per minute.

    (4) Fuel oil must be discharged in less than 25 strokes.

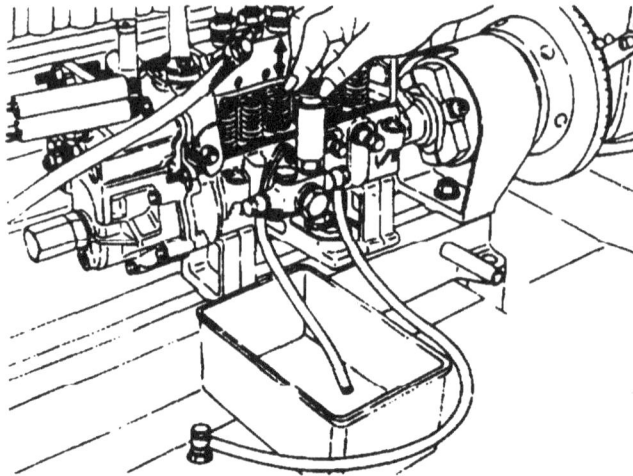

FIGURE 3-33. Priming Pump Test

aj. Perform feed pump capacity test as follows:

NOTE

Feed pump can be tested for capacity while installed on engine. Discon-nect discharge hose and insert in measuring container. Crank engine for 15 seconds. Supply rate should be at least 60 cc. Replace feed pump if it fails to meet supply rate.

(1) Pump should be piped as in step ah, except place end of discharge pipe in a measuring cylinder, refer to FIGURE 3-34.

(2) Operate injection pump at speeds specified in TABLE 3-2 and check for corresponding supply rate.

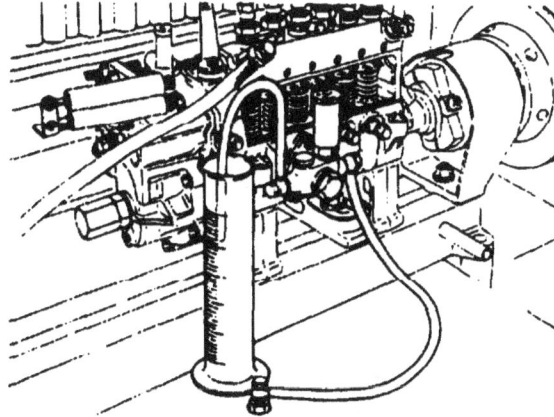

FIGURE 3-34. Capacity lest

TABLE 3-2. Capacity Test

Pump Speed	Supply Rate
100 rpm	More than 60 cc/15 seconds
500 rpm	More than 340 cc/15 seconds
1000 rpm	More than 405 cc/15 seconds
1500 rpm	More than 430 cc/15 seconds

ak. Perform feed pump feed pressure measurement as follows:

(1) Pump suction as piped in step ah, refer to FIGURE 3-35.

(2) Pump discharge connected to test stand pressure gage

(3) Operate injection pump at 600 rpm.

(4) Ensure that feed pump maintains a feed pressure of 25.6 to 35.5 psi (176.5 to 244.8 kPa) al.

Perform feed pump air tightness test as follows:

(1) Remove feed pump from pump housing. Refer to paragraph 3-11.2.2, step h.

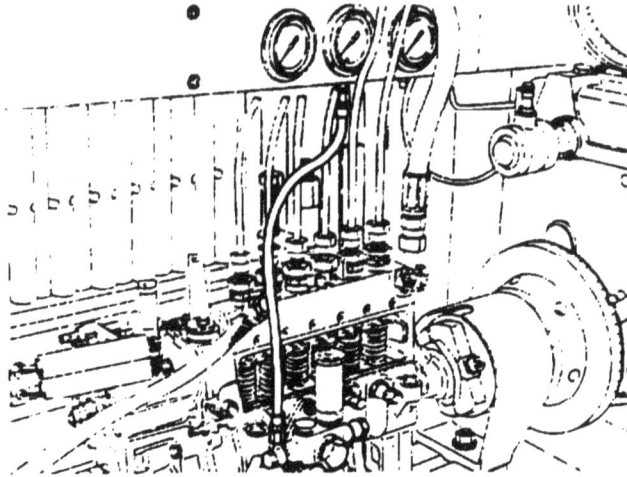

FIGURE 3-35. Feed Pump Pressure Test

(2) Install plug in discharge port.

(3) Connect an air supply to suction port and apply a pressure of 28.5 psi (196.5 kPa).

(4) Immerse feed pump in a container of diesel fuel and check for air leakage. There must be no air leakage.

(5) Install feed pump on pump housing. Refer to paragraph 3-11.4.2.

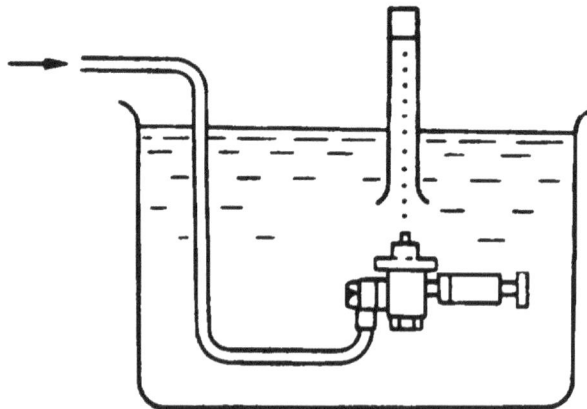

FIGURE 3-36. Air Tightness Test

am. Remove coupling from camshaft.

an. Install cover (13, FIGURE 3-8) with new gasket (14) and secure with two new gaskets (12) and screws (11).

ao. Position bracket (8) and new gasket (10) on drive end of housing and secure with four washers (7) new lock-washers (6), and nuts (5)

ap. Install timer assembly on camshaft and secure with new lockwasher (6, FIGURE 3-7) and nut (5). Attach spanner (157916-5320) to timer assembly to hold it while tightening nut (5) to 45 ft-lbs (61 Nm)

3-11.6. Installation.

NOTE

If a replacement injection pump does not include a mechanical gover-nor, skip steps a thru g and proceed with step h.

a. Disconnect and remove spring (1, FIGURE 3-37).

b. Remove stud (2) and washer (3).

c. Position pump in a pan to collect any oil which may be contained in governor housing.

d. Remove two screws (4) and four screws (5).

e. Partially separate cover (6) from housing. While holding cover in place with one hand, disconnect spring (8) and link (9) from inside housing. This will allow separating cover (6) from housing.

f. Remove gasket (10).

g. Using a hammer and screwdriver, loosen and remove nut (11) and washer (12). Remove flyweight assembly (13).

Position fuel injection pump and new packing (4, FIGURE 3-7) on engine front plate, aligning pump gear setting mark 'Z' with camshaft gear setting mark 'z-z'. Secure pump with six bolts (1 and 2), tighten bolts 168 in-lbs (18.9 Nm).

i. Install engine front cover access plate, refer to paragraph 3-25.3 step

j. Connect/assemble engine speed control system to fuel injection pump. Refer to end item maintenance manual.

k. Position oil feed tube to side of pump and secure with two new preformed packings (4, FIGURE 3-8) and fluid passage bolt (3). Tighten bolt to 132 in-lbs (14.9 Nm).

l. Remove caps and position fuel tube from fuel filter on top front of fuel injection pump and secure with two new seals (2) and fluid passage bolt (1). Tighten bolt to 132 in-lbs (14.9 Nm)

m. Remove caps and connect fuel tubes to top and bottom of fuel feed pump (10, FIGURE 3-7).

n. Remove caps and connect fuel tubes to delivery valve ports on fuel injection pump, refer to paragraph 3-10.7, step c.

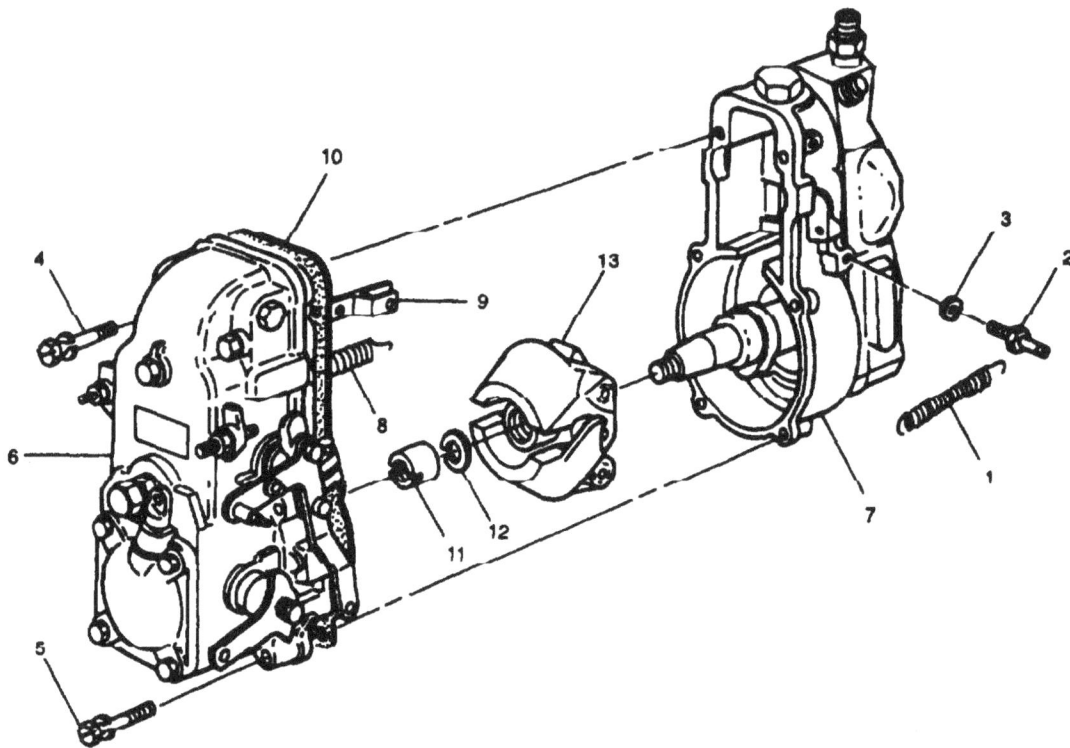

FIGURE 3-37. Removal of Mechanical Governor

o. Service engine cooling and lubrication systems, refer to end item maintenance manual.

p. Bleed air from fuel system, refer to paragraph 3-12.

q. Operate engine and check for leaks. 3-12._

BLEEDING AND PRIMING FUEL SYSTEM.

3-12.1.

a. Loosen priming pump cap on fuel injection pump.

b. Prime fuel injection pump several times.

c. Loosen bleeder screw (40, FIGURE 3-8) on fuel injection pump.

WARNING

Fuel is flammable. Keep fuel away from heat and open flame. Failure to observe this warning could result in severe personal injury or death.

NOTE

Drain fuel into suitable container.

d. Operate priming pump (20, Figure 3-7) until no more bubbles are visible in fuel being discharged from bleeder screws.

e. Retighten bleeder screws and priming pump cap securely after bleeding.

3-13. FUEL INJECTION PUMP TIMING.

3-13.1. Checking Procedure.

a. Check that notched line on injection pump fixing flange is aligned with injection pump mounting flange's notched line.

b. Align crankshaft pulley TDC mark with pointer.

c. Rotate crankshaft pulley counterclockwise approximately 30 degrees.

d. Disconnect injection pipe from delivery valve No. 1.

e. Remove No. 1 plate assembly (32, FIGURE 3-8), valve body (33), spring (34); gasket (36). and valve (35).

f. Install valve body (33) and tighten to 31 ft-lbs (42.0 Nm).

CAUTION

Take care to avoid entry of dirt or foreign particles in fuel injection pump while delivery valve is removed. Otherwise, damage to pump could oc-cur.

---

g. Hold fuel control lever to full fuel position.

h. Slowly turn crankshaft pulley clockwise and at same time operate priming pump to cause fuel to flow from No. 1 delivery valve.

i. When fuel stops flowing from No. 1 delivery valve, stop pumping and stop rotating crankshaft. This is correct fuel injection timing.

j. Observe and ensure that crankshaft pulley middle timing mark (18 degrees) is aligned with pulley.

k. Blow remaining fuel from No. 1 delivery valve holder and check that no fuel flows as pumping is continued.

l. Remove delivery valve body (33). Install valve (35), gasket (38), spring (34), valve body (33), and plate assembly (32), refer to paragraph 3-11.4.3, steps c, d, and af.

m. Reconnect injection pipe to No. 1 delivery valve, refer to paragraph 3-10.7, step c.

CAUTION

Operating engine with priming pump cap in unscrewed position will result in priming pump failure.

---

n. Tighten priming pump cap.

3-13.2. <u>Adjustment Procedure.</u>

  a. Align timing pointer at timing gear case and middle (18 degrees) injection timing notch line on crankshaft pulley.

  b. Disconnect injection pipe from delivery valve No. 1. Cap openings.

  c. Remove No. 1 plate assembly (32, FIGURE 3-8), valve body (33), spring (34), gasket (36), and valve (35). refer to paragraph 3-11.2.2.

  d. Install valve body (33) and tighten to 31 ft-lbs (42.0 Nm).

CAUTION

Take care to avoid entry of dirt or foreign particles in fuel injection pump while delivery valve is removed. Otherwise, damage to pump could occur.

  e. Hold fuel control lever to full fuel position.

NOTE

Injection timing adjustment is done by pivoting injection pump at drive shaft. Pivot pump body toward outside of engine to advance timing. Piv-ot pump body toward inside of engine to retard timing.

  f. Disconnect other injection pipes at delivery valves to allow pump to pivot easily. Cap openings.

  g. Loosen four nuts (5) mounting pump and pivot pump body.

NOTE

Movement of 0.04 in. (1.0 mm) corresponds to approximately 2 degrees in crank angle.

  h. Operate priming pump to allow fuel to flow from No. 1 delivery valve.

i. While continuing to pump as described in step h, pivot pump body toward inside or outside of engine until fuel stops flowing from No. 1 delivery valve. This is the correct fuel injection timing.

j. Blow remaining fuel from No. 1 delivery valve holder and check that no fuel flows as pumping is continued.

NOTE

Remove fuel priming pump and PCV hoses as necessary to torque nuts.

k. Tighten four nuts (5) securing pump to 168 in-lbs (19.0 Nm).

l. Remove delivery valve body (33). Install valve (35), gasket (36), spring (34), and valve body (33), refer to paragraph 3-11.4.2.

m. Remove caps and reconnect injection pipes to delivery valves, refer to paragraph 3-10.7.

```
┌─────────────────┐
│     CAUTION     │
└─────────────────┘
```

Operating engine with priming pump cap in unscrewed position will result in priming pump failure.

SECTION VI. LUBRICATION SYSTEM

3-14.<u>OIL FILTER.</u>

3-14.1.<u> Cartridge Replacement.</u>

    a. Drain lubrication system, refer to end item maintenance manual.

    b. Loosen cartridge (1, FIGURE 3-38) by turning counterclockwise with a filter wrench. c.

Discard used cartridge.

    c. Wipe filter head (2) with a clean rag.

    e.Apply light coat of oil to new cartridge (1) gasket.

    f. Place cartridge on filter head (2) and turn clockwise until cartridge is hand-tight on head.

    g. Using filter wrench, turn cartridge an additional 1 to 1-1/4 turns.

    h. Service lubrication system, refer to end item maintenance manual.

    i. Operate engine and check for leakage.

3-14.2.<u> Removal.</u>

    a. Drain lubrication system, refer to end item maintenance manual.

    b. Remove four screws (3, FIGURE 3-38) securing filter head (2) to engine; remove filter head and o-ring (4).

NOTE

Cartridge (1) does not have to to be removed to remove head.

    c. Cover opening to prevent entry of dirt.

3-14.3.<u> Installation.</u>

    a. Wipe engine and filter head (2, FIGURE 3-38) mating areas with a clean rag.

    b. Install o-ring (4). Position head (2) on engine and secure with four screws (3). Tighten screws to

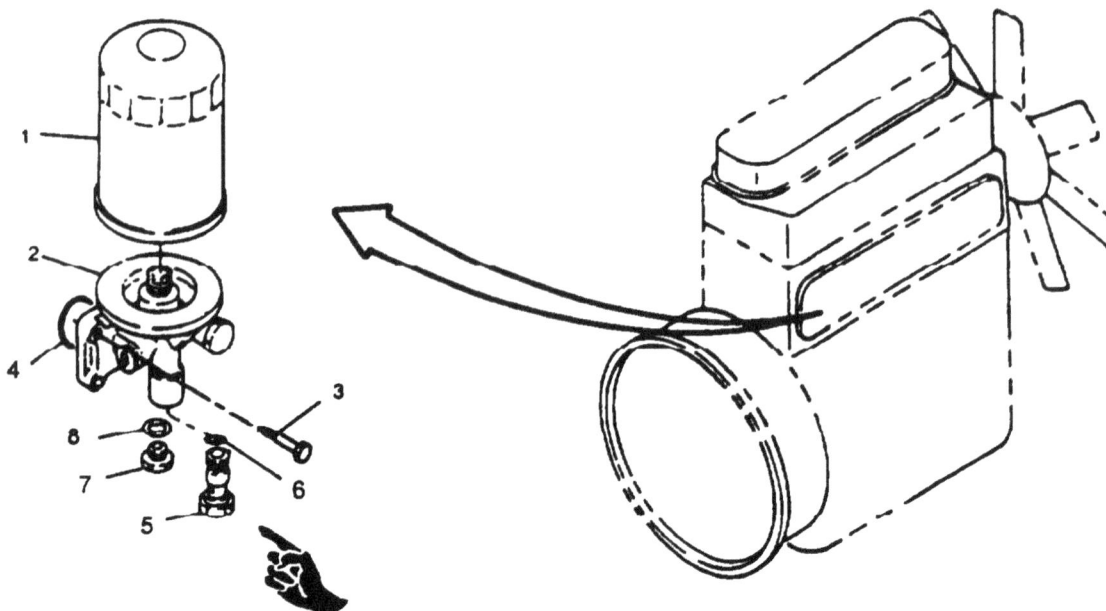

FIGURE 3-38. Oil Filter

c. Install cartridge (1) if necessary, refer to paragraph 3-14.1.

d. Service lubrication system, refer to end item maintenance manual.

e. Operate engine and check for leakage. 3-15.

OIL PRESSURE RELIEF VALVE. 3-15.1. Removal.

a. Drain lubrication system, refer to end item maintenance manual.

b. Remove oil pressure relief valve (5, FIGURE 3-38) and o-ring (6).

c. If necessary, remove plug (7) and seal (8). Discard seal (8).

3-15.2. Inspection.

a. Inspect oil pressure relief valve (5, FIGURE 3-38) for damaged threads or other damage.

b. Inspect plug (7) for damage.

3-15.3. Testing.

    a. Tag, disconnect, and remove oil pressure sending unit, if supplied.

    b.   Install oil pressure gage (0-150 psi).

    c.   Start engine and check oil pressure relief valve opening. Valve should open between 82.5 to
88.2

         psi (568 to 608 kPa).

    d. Replace oil pressure relief valve if valve fails to open as specified above.

    e.   Remove oil pressure gage and install oil pressure sending unit (if supplied). Connect electrical connection and remove tag.

3-15.4. Installation.

    a.   If removed, install plug (7) with new seal (8). Torque plug between 18.1 to 25.3 ft-lbs (24.5 to 34.3 Nm).

    b.   Install oil pressure relief valve (5) and o-ring (6). Torque relief valve between 18.1 to 25.3 ft-lbs (24.5 to 34.3 Nm)

    c.   Service lubrication system, refer to end item maintenance manual.

    d. Operate engine and check for leakage.

3-16. OIL PIPING.

3-16.1. Removal.

    a. Drain lubrication system, refer to end item maintenance manual.

    b.   Remove fluid passage bolt (1, FIGURE 3-39) securing valve rocker oil feed pipe (3). Remove
pipe   and two gaskets (2). Discard gaskets (2).

    c.   Remove two fluid passage bolts (4) and bolt (6) securing fuel injection pump oil pipe (7). Re-

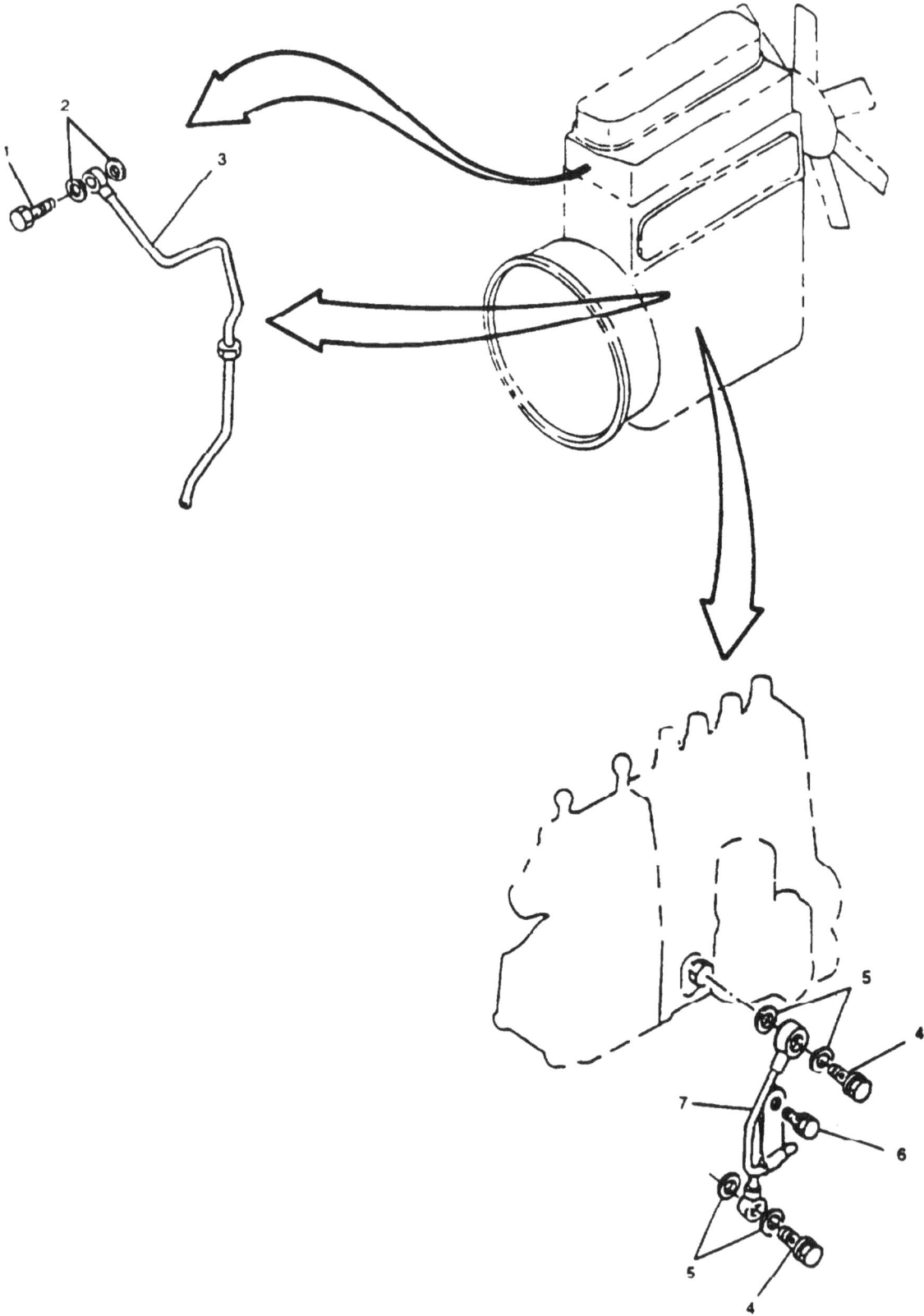

FIGURE 3-39. Oil Piping

3-16.2. Installation.

a.Position fuel injection pump oil pipe (7, FIGURE 3-39) and secure with four new gaskets (5), two fluid passage bolts (4), and bolt (6). Tighten bolts (4) to 84 in-lbs (9.5 Nm).

b.Position valve rocker oil fuel pipe (3) and secure with two new gaskets (2) and fluid passage bolt (1). Tighten bolt to 84 in-lbs (9.5 Nm).

c.Service lubrication system, refer to end item maintenance manual. d.Oper-

ate engine and check for leaks.

3-17.   OIL PAN.

3-17.1. Removal,

a.Drain lubrication system, refer to end item maintenance manual.

b. Disconnect PVC hose.

c.Remove dipstick (I, FIGURE 340) and gasket (2). Discard gasket. d.Re-

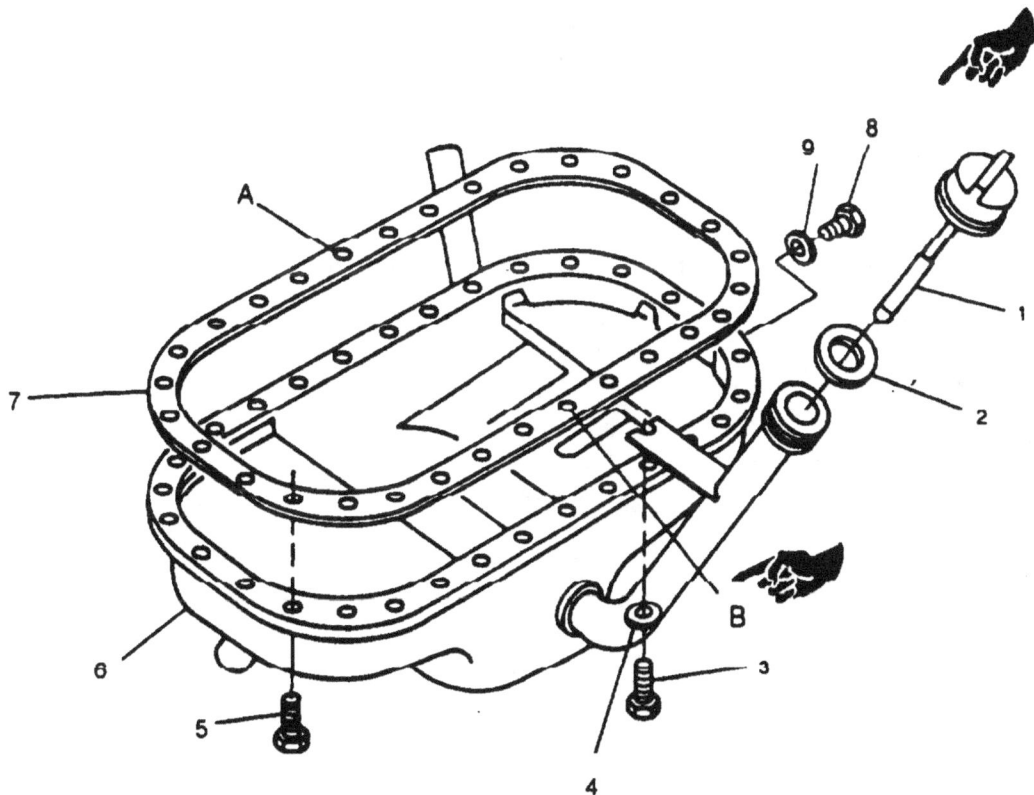

FIGURE 3-40. Oil Pan Assembly

e. Support oil pan and remove thirty screws (5) securing oil pan. Remove oil pan (6) and gasket (6). Discard gasket.

f.   If not removed, remove drain plug (8) and gasket (9).

3-17.2. Inspection.

a. Inspect dipstick (1, FIGURE 3-40) for legibility. Replace if damaged.

b. Inspect gasket (2) for deterioration and replace as necessary.

c. Inspect oil pan (6 for cracks, dents, or other damage. Replace if damaged.

3-17.3. Installation.

a.   Wipe crankcase flange and oil pan flange and remove any traces of oil pan gasket or liquid gasket.

Apply an even coat of sealing compound (FORMAGASKET2) to crankcase flange.

b.
Affix new gasket (7, FIGURE 3-40) to crankcase flange.

c.
Remove screw (3) and washer (4) securing tube.

d.
Position oil pan (6) and install all thirty screws (5) hand-tight. Tighten screws to 72 in-lbs (8.1 Nm)
e.   starting at point 'A' and 'B' alternately and work in a counterclockwise direction from each of these points until all thirty screws are tightened.

Install screw (3) and washer (4) secunig tube. Tighten to 168 in-lbs (19 Nm).

f.
Install oil drain plug (8) with new gasket (9).

g.   Install dipstick (1) and new gasket (2).

h.
Connect PCV hose.

i.
Service lubrication system, refer to end item maintenance manual.

j.   Operate engine and check for leaks.

k.

3-18. OIL PUMP.

3-18.1. Removal / Disassembly.

    a.Remove oil pan, refer to paragraph 3-17.1. b.Remove

crankcase, refer to paragraph 3-29.1. c.Disconnect oil pipe (1,

FIGURE 3-41) at sleeve nut.

    d.Remove two screws (2) securing pump assembly and remove pump.

    e.Disconnect pipe (1) from pump.

    f.   Unscrew strainer assembly (3) from pump cover (5).

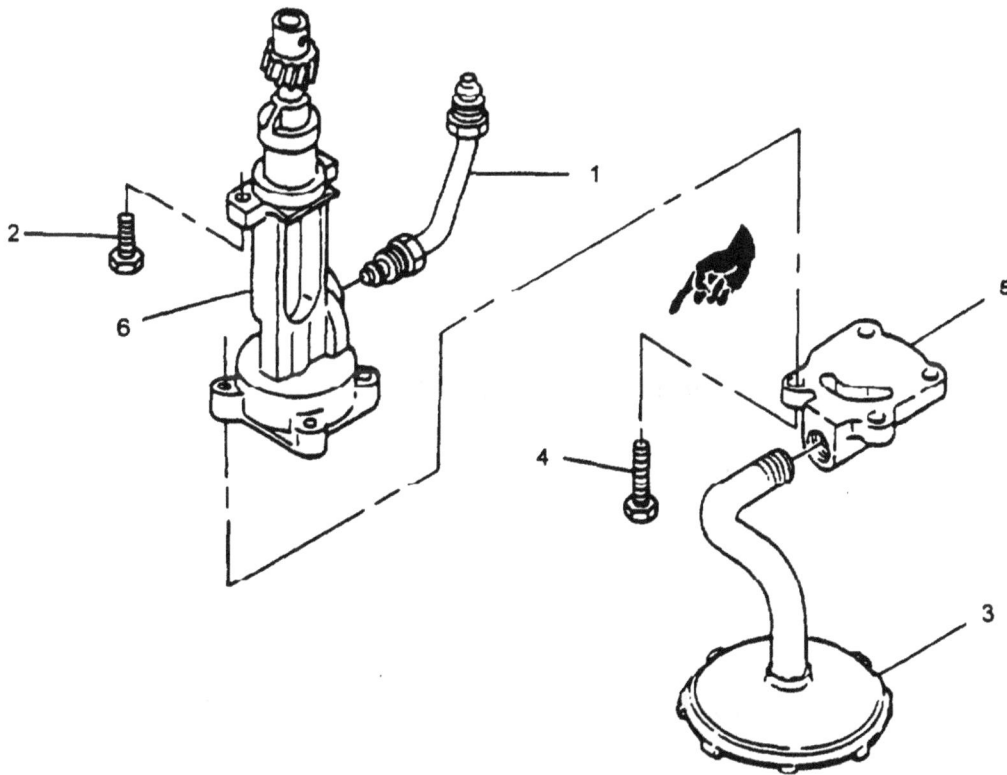

FIGURE 3-41. Oil Pump Assembly

3-18.2. Inspect and Measure.

    a. Using a feeler gage, measure clearance between vane, rotor, and cover as shown in View A, FIGURE 3-42. Standard clearance is 0.0008 to 0.0028 in. (0.02 to 0.07 mm) and limit is 0.0059 in. (0.15 mm) If clearance exceeds limit, replace pump rotor (8, FIGURE 3-41).

    b. Using a feeler gage, measure clearance between rotor and vane as shown in View B, FIGURE 3-42. Standard clearance is 0.0055 in. (0.14 mm) or less and limit is 0.0079 in. (0.20 mm). If clearance exceeds limit, replace pump rotor (8, FIGURE 341).

    c. Using a feeler gage, measure clearance between vane and pump body as shown in View C, FIGURE 3-42. Standard clearance is 0.0079 to 0.0106 in. (0.20 to 0.27 mm) and limit is 0.016 in. (0.40 mm). If clearance exceeds limit, replace pump rotor (8, FIGURE 341).

    d. Inspect strainer (3) for blockage and clean as necessary.

    e. Inspect all parts for excessive wear or damage and replace defective components. 3-18.3.

Assembly/Installation.

    a. Position pump cover (5, FIGURE 3-41) on pump rotor (6) and secure with four screws (4).

    b. Attach stainer assembly (3) to pump cover (5).

    c. Attach oil pipe (1) to pump.

    d. Position pump in engine and secure with two screws (2). Tighten screws to 168 in-lbs (19.0 Nm).

    e. Connect other end of oil pipe (1) to engine at sleeve nut.

    f. Install crankcase, refer to paragraph 3-29.3.

    g. Install oil pan, refer to paragraph 3-17.3. 3-19.

OIL PRESSURE TEST.

    h. Attach pressure gage at pressure port shown on FIGURE 3-43.

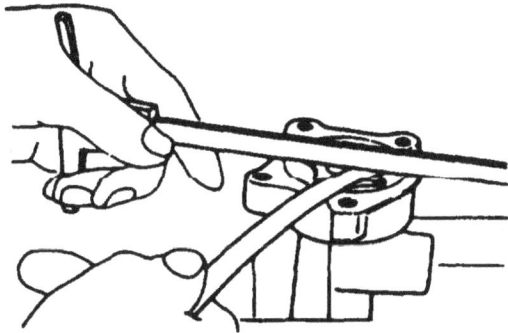

VIEW A

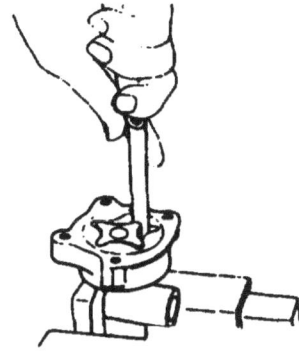

VIEW B

VIEW C

FIGURE 3-42. Oil Pump Inspections

```
CAUTION
```

Before checking the oil pressure, warm up engine to allow the lubricat-ing oil
to reach operating temperature or high oil pressure readings will occur.

b. At 850 rpm engine speed and 200°F (93°C) operating temperature, gage should show a minimum pressure of 14 psi (100 kPa).

c. At rated speed (1800 to 2500 rpm) and 220°F (105°C) operating temperature, gage should show a pressure between 40 and 70 psi (277 to 483 kPa).

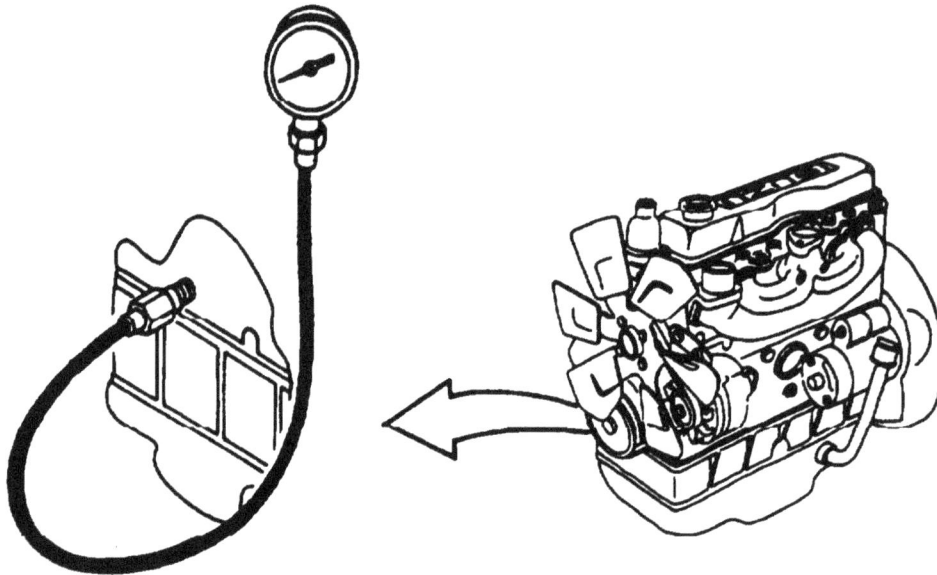

FIGURE 3-43. Oil Pressure Test Setup

SECTION VII. INTAKE AND EXHAUST SYSTEM MAINTENANCE

3-20. PCV ASSEMBLY.

a.   Remove two clamps (1 and 2, FIGURE 3-44) attaching hose (3) to intake manifold and PCV valve (14); remove hose.

b.   Remove two clamps (4) attaching hose (5) to PCV valve (14) and breather pipe (17); remove hose.

c.   Remove two clamps (6) attaching hose (8) to PCV valve (14) and check valve (10). Remove screw securing loop clamp (7); remove hose (8).

d.   Remove hose clamp (6) attaching hose (9) to check valve (10); remove check valve (10).

e.   Remove hose clamp (6) attaching other end of hose (9) to tube (25); remove hose (9).

f.   Remove two bolts (11), lockwashers (13) and washers (12) securing PCV valve (14); remove valve. Discard lockwashers (13).

g.   Remove two screws (15) securing breather pipe (17) to tappet cover (28). Remove bracket (16) breather pipe (17) gasket (18) and baffle (19). Discard gasket (18)

h.   Remove two clamps (20) attaching hose (21) to tappet cover (28) and tube (25); remove hose.

i.   Remove two clamps (20) attaching hose (22) to tube (25) and oil pan; remove hose.

j.   Remove bolt (23) and lockwasher (24) securing tube (25); remove tube. Discard lockwasher (24).

k.   Remove twelve screws (33) and two screws (26) securing tappet cover (28). Remove two clips (27) cover (28) and gasket (29). Discard gasket.

l.   Remove bolt (30) and lockwashers (31) securing baffle (32); remove baffle. Discard lockwasher.

3-20.2. Inspection.

a. Inspect hoses for deterioration, cracks, cuts, or other damage. Replace if defective.

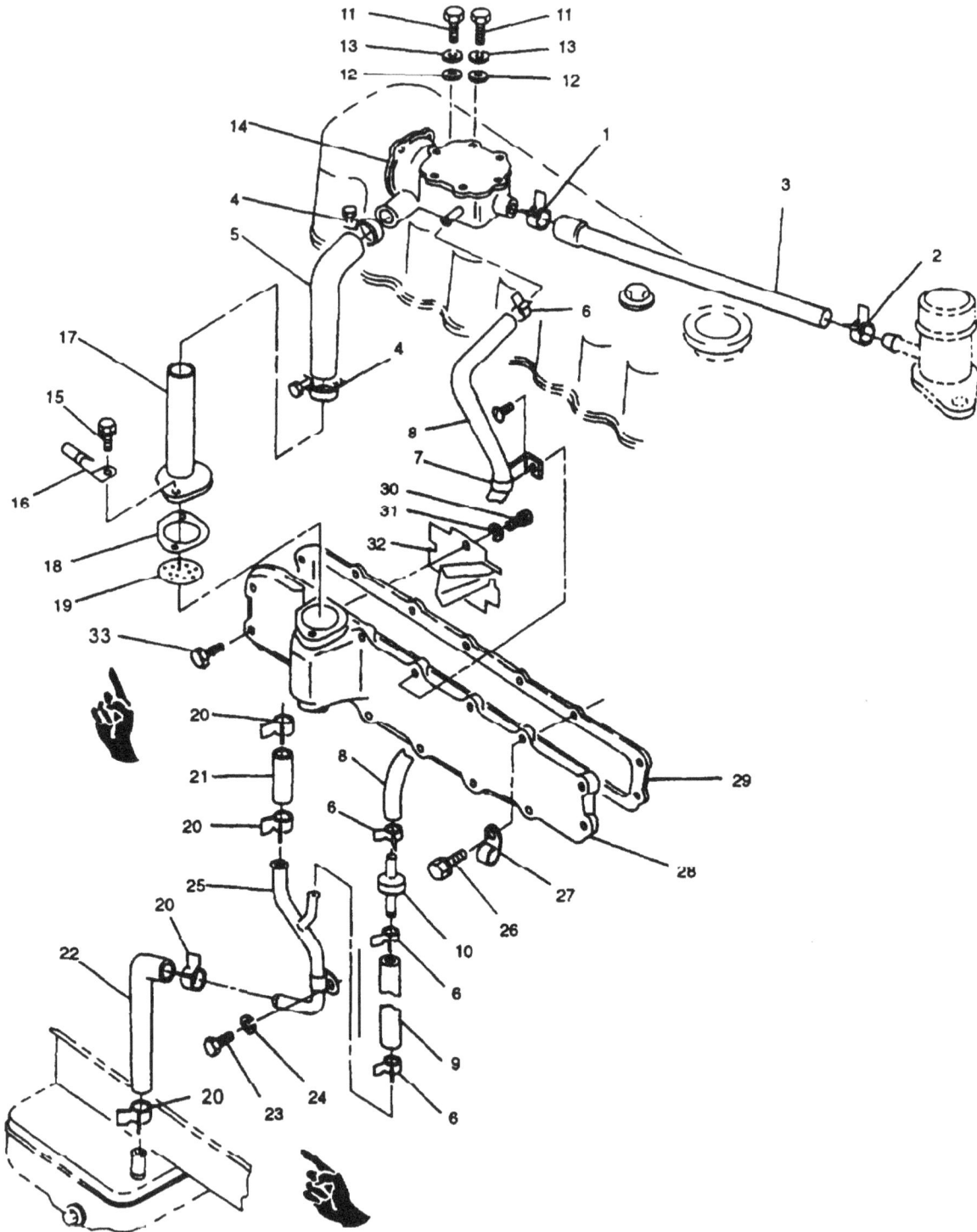

FIGURE 3-44. PCV Assembly

b.Inspect tubes for deformity, cracks, or other damage. **Replace** if defective.

c.Inspect clamps for deformity, weakness, or other damage. Replace if defective.

d.Inspect check valve for damage. Blow into check valve. At **one** end, **air** should pass freely. At other end, air will not pass through.

3-20.3. Installation.

a. Position baffle (32, FIGURE 344) and secure with bolt (30) and new lockwasher (31).

b. Position new gasket (29) tappet cover (28), and two clips (27) on block and secure with twelve screws (33) and two screws (26) hand tightened. Starting with top center screws and working alternately top to bottom toward both ends, tighten screws evenly to 168 in-lbs (19.0 Nm).

c. Position tube (25) on side of engine and secure with bolt (23) and new lockwasher (24).

d. Attach hose (22) to tube (25) and oil pan and secure with two clamps (20).

e. Attach hose (21) to tappet cover (28) and tube (25) and secure with two clamps (20).

f. Position baffle (19) new gasket (18), breather pipe (17) and bracket (16) on tappet cover (28) and secure with two screws (15).

g. Position PCV valve (14) on engine and secure with two bolts (11), new lockwashers (13) and washers (12).

h. Attach one end of hose (9) to tube (25) and secure with clamp (6).

i. Install check valve (10) in other end of hose (9) and secure with clamp (6).

j. **Attach** hose (8) to PCV valve (14) and check valve (10) with two clamps (6). Secure loop clamp (7) to engine with screw.

k. Attach hose (5) to PCV valve (14) and breather pipe (17) with two clamps (4).

l. Attach hose (3) to PCV valve (14) and intake manifold with clamps (1) and (2).

3-21. INTAKE AND EXHAUST MANIFOLDS.

3-21.1. Removal.

a. Remove air intake and exhaust piping, refer to end item maintenance manual.

NOTE

The intake and exhaust manifold must be removed as an assembly.

b. Remove two bolts (3, FIGURE 3-45) seven nuts (4) lockwashers (5) and washers (6) securing intake manifold (7) and exhaust manifold (8).

c. Remove two spacers and angle bracket, refer to end item maintenance manual.

d. Remove clamp (1) and hose (2) to intake manifold.

e. Remove intake manifold (7) exhaust manifold (8) and gasket (9). Discard lockwashers (5) and gasket (14).

f. Remove two nuts (10), lockwashers (11), and washers (12) securing inlet pipe (13) to intake manifold (5). Remove intake pipe and gasket (14). Discard lockwashers (11) and gasket (14).

g. If damaged, remove studs (15 and 16).

3-21.2. Inspection.

a. Inspect manifolds for cracks, corrosion, or other damage. Replace if damaged.

b. Inspect attaching hardware (bolts, studs, and nuts) for thread damage. Replace if damaged.

3-21.3. Installation.

a. If removed, install studs (15 and 16, FIGURE 345).

b. Position new gasket (14) and inlet pipe (13) on intake manifold (7) and secure with two nuts (10), new lo&washers (11) and washers (12).

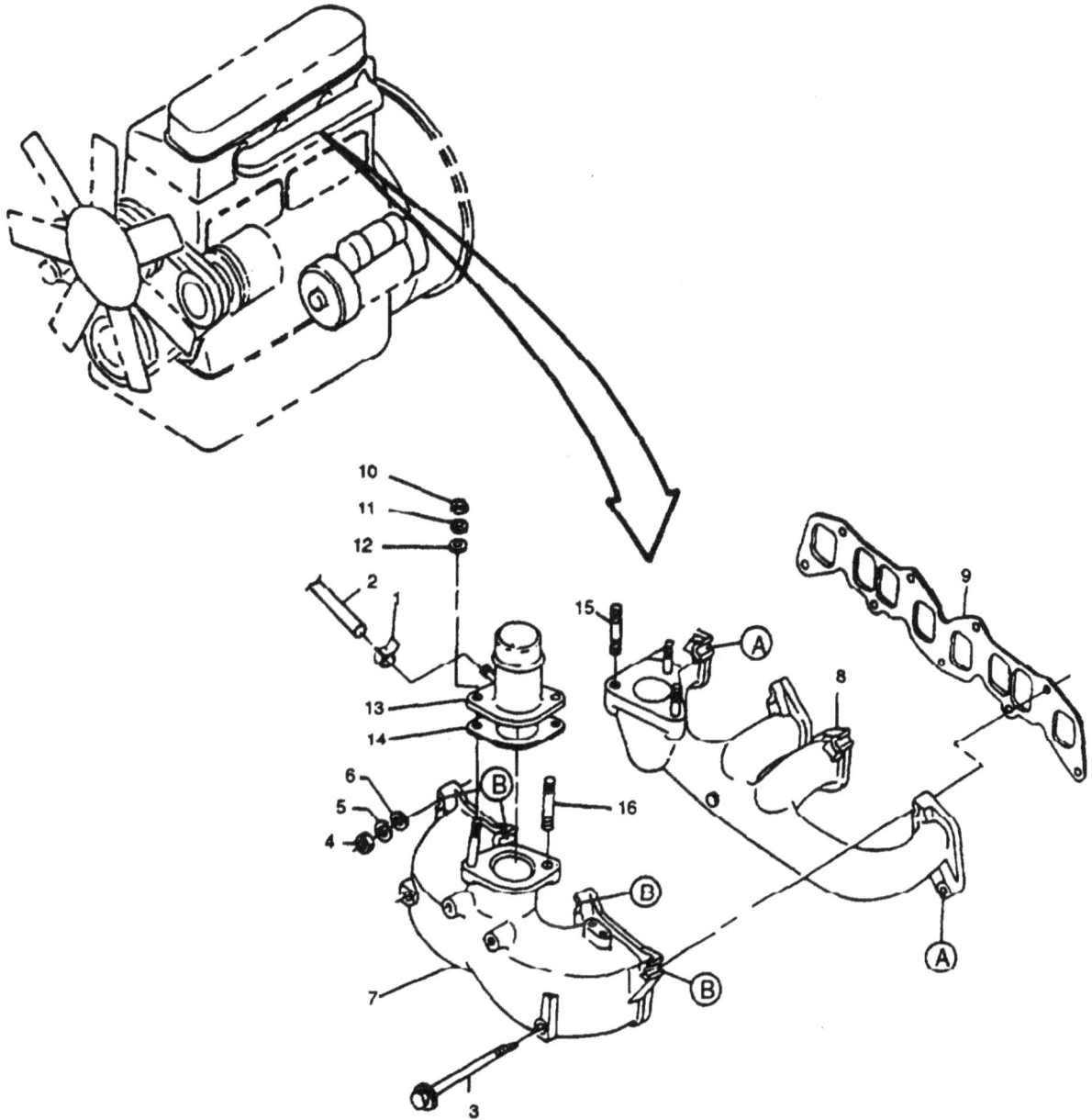

FIGURE 3-45. Intake and Exhaust Manifolds

c. Clean intake manifold (7) and exhaust manifold (8), mating surfaces, and cylinder head mating surface of any traces of old gasket.

d. Position new gasket (9) on studs in cylinder head.

NOTE

Intake manifold and exhaust manifold must be installed as an assembly.

e. Position exhaust manifold (8) and intake manifold (7) on studs and secure hand-tight at points 'A' and 'B' with nuts (4), new lockwashers (5), and washers (6).

f. Install two, bolts (3) hand-tight.

g. Tighten all nuts (4) and bolts (3) to 168 in-lbs (19.0 Nm).

h. Install hose (2) to inlet pipe (13) and secure with clamp (1).

SECTION VIII. ELECTRICAL SYSTEM

3-22. GLOW PLUGS. 3-22.1.

Removal.

    a. Loosen nuts on glow plugs and remove connector (1, FIGURE 3-46).

    b. Remove glow plugs (2).

3-22.2. Inspection.

    a. Inspect electrical leads for cracks, deterioration, cuts, corrosion, or other damage. Replace if damaged.

    b. Inspect glow plugs for damage to threads or corrosion. Replace if damaged.

NOTE

Always replace all four glow plugs as a set. Never replace just one glow plug.

3-22.3. Testing.

    a. Disconnect connector (1) from glow plug.

    b. Set multimeter for ohms and connect one lead of multimeter to glow plug terminal and other lead to body.

    c. Normal resistance of glow plug is 4.4 to 4.8 ohms. If reading is zero, glow plug is shorted. If reading is high, glow plug is open. If shorted or open, replace glow plug.

3-22.4. Installation.

    a. Install four glow plugs (2, FIGURE 3-46) in cylinder head. Tighten to 168 in-lbs (19.9 Nm).

    b. Install connector (1) by tightening nuts on glow plugs. 3-23.

STARTER ASSEMBLY. 3-23.1.Removal.

    a. Tag and disconnect electrical leads from starter.

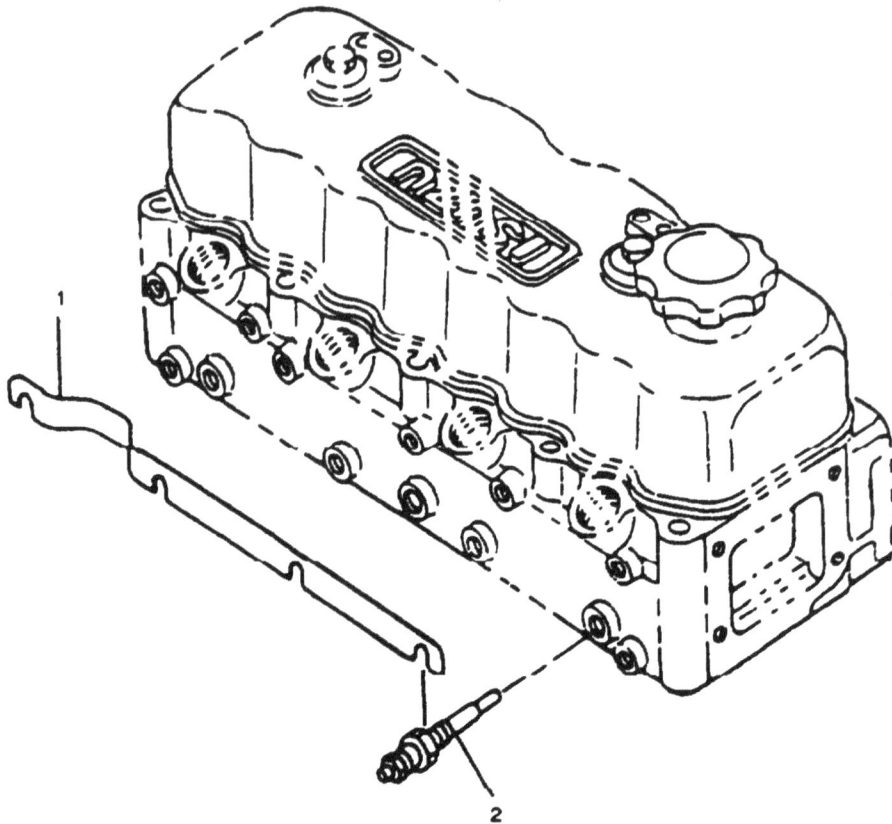

FIGURE 3-46. Glow Plugs Assembly

b. Support weight of starter and remove two nuts (1, FIGURE 3-47) lockwashers (2) and washers (3). Remove starter from flywheel housing.

c. Cover opening in flywheel.

3-23.2. Disassembly.

a. Benchmark gear case (23, FIGURE 3-47), field coil (24) and housing (11) to aid in aligning parts during assembly.

b. Disconnect field coil connecting wire from terminal on magnetic switch (5).

c. Remove two bolts (4) securing magnetic switch (5) to gear case (23); remove magnetic switch.

d. Remove two screws (6).

e. Remove two bolts (9) and washers (10).

NOTE

If housing (11) is difficult to separate from field coil (24) tap lightly with plastic mallet.

f. Remove housing (11) from field coil (24).

g. Remove positive brushes (12) (attached to field coil) from brush holder (15) by raising brush spring (13) and removing brush from holder.

h. Raise springs (13) on negative brushes (14). Raise negative brushes part way off commutator and hold in this position by setting spring against brush.

i. Remove brusher holder (15).

NOTE

Note installation of shift fork (20) to aid in assembly.

j. Remove nut (16) lockwasher (17) and stud (19) securing shift fork (20) in gear case (23).

k. Remove dust covers (21 and 22).

l. Remove gear case (23).

m. Remove shift fork (20).

n. Remove field coil (24).

NOTE

To remove pinion stopper (26) move stopper toward armature to expose retainer ring.

o. Remove washer (25) pinion stopper (26) and clutch (27) from armature shaft (18).

p. Remove bracket (28) with center metal (29) from armature shaft.

NOTE

Note from which end of shaft washers (7 and 30) come.

q. Remove thrust washers (7 and 30) from armature shaft.

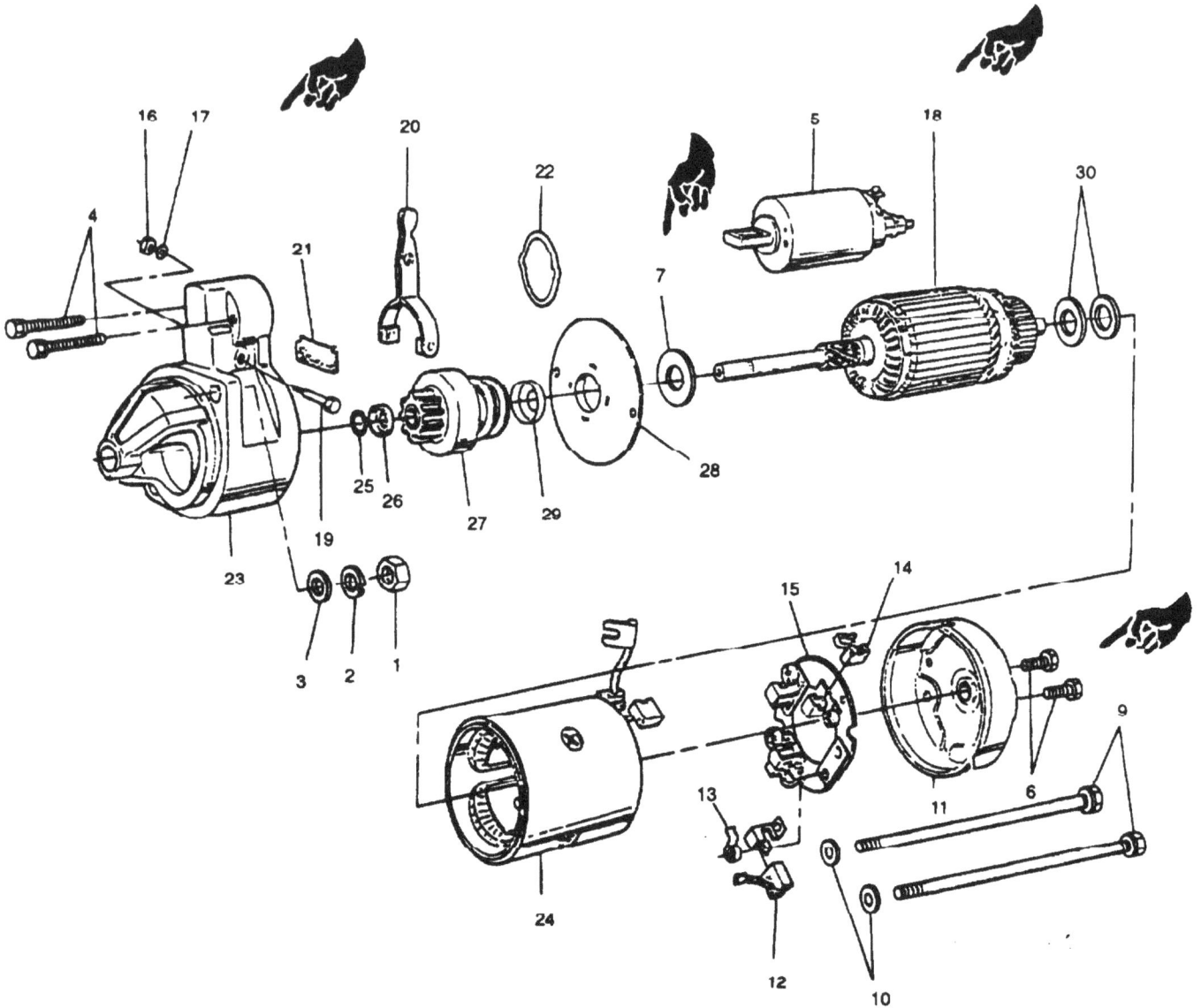

FIGURE 3-47. Starter Assembly

3-23.3. Testing.

3-23.3.1. Testing Starter (Installed).

    a. Make sure batteries are fully charged and that all battery and starter cables are serviceable and properly installed.

b. Set multimeter for volts DC and as shown in FIGURE 3-48, Test A. If voltage is indicated, solenoid is defective.

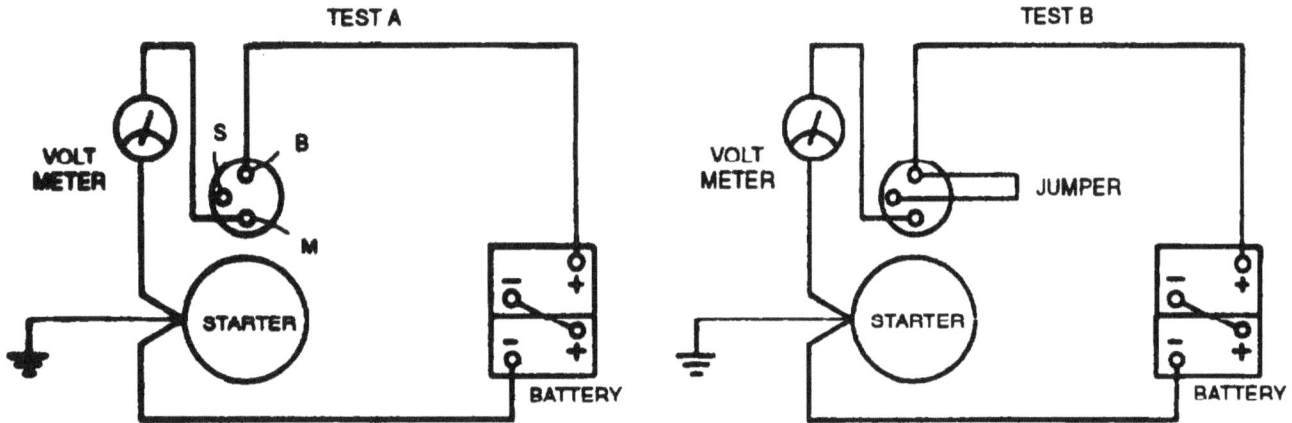

TEST A   TEST B

FIGURE 3-48. Starter Solenoid Test Circuit

c. Momentarily connect a jumper as shown in FIGURE 3-48, Test B. Multimeter should indicate battery voltage and starter should crank the engine. If multimeter does not read battery voltage, the solenoid is defective. If multimeter indicates battery voltage, but starter does not operate, starter is defective.

3-23.5.2. Testing Starter Components.

    a. Test armature for grounds.

      (1) Bet multimeter for ohms and touch armature shaft and each commutator bar with multimeter leads.

      (2) If multimeter indicates a low reading (approximately zero ohms), armature is grounded and requires replacing.

    b. Test armature for open circuits.

      (1) Check for continuity between commutator segments with multimeter.

      (2) If an open is indicated, armature must be replaced.

c. Teat field ceil.

(1) Check for continuity between brushes with a multimeter.

(2) If there is no continuity indicated, field coil is open and field coil frame must be replaced.

(3) Check for continuity between field coil frame and brushes.

(4) If continuity is indicated, field coil is grounded and field coil frame must be replaced.

d. Test solenoid.

(1) Check for continuity between solenoid terminals M and B with a multimeter. Refer to FIGURE 3-48.

(2) No continuity should be indicated when solenoid plunger is in normal (out) position.

(3) Push solenoid plunger in and note multimeter. Multimeter should indicate continuity. If no continuity is indicated, solenoid must be replaced.

3-23.4. Cleaning and Inspection.

| WARNING |

Cleaning solvent is flammable and toxic to eyes, akin, and respiratory tract. Skin/eye protection required. Avoid repeated/prolonged contact.

| WARNING |

Good general ventilation is normally adequate.
Compressed air used for cleaning can create airborne particles that may

CAUTION

Do not clean overrunning clutch in solvent or other cleaning solution.
Washing clutch will remove grease which may result in premature fail-ure
of the clutch.

a. b. Wipe all metallic parts with a cleaning cloth (TX-1250) that has been slightly dampened with cleaning solvent.

Inspect all parts for damaged threads, cracks, distortion, or other visible damage.

c. Inspect armature commutator. If commutator is dirty or discolored, clean with abrasive paper (#400). Use compressed air to blow sand out from between commutator segments.

d. Inspect commutator for pits, scoring, excessive carbon, and oil. Turn armature commutator in a lathe to re-move surface pits, scoring, and contamination. Normal diameter is 1.89 in. (48.0 mm), minimum after turning is 1.77 in. (45.0 mm). Maximum runout should be 0.016 in. (0.4 mm) Total Indicator Reading (TIR).

e. Clean around brushes and holders by wiping off all brush dust and dirt. If brushes are shorter than 0.45 in. (11.5 mm) they should be replaced.

f. Check for free movement of brushes. Brushes should move freely when placed in brush holders. Replace brush springs if weak or worn.

g. Inspect brush holders for discoloration, burrs, or signs of high temperature indicating defective insulators. Replace brush holder.

h. Inspect armature shaft for damage and wear. Replace armature if damaged.

NOTE

If pinion gear is worn or damaged, inspect flywheel ring gear also.

i. Inspect armature shaft gear, bearings, and pinion gear for wear or damage. Replace any part that is damaged.

j. When pinion gear/clutch is placed upon armature shaft, pinion gear should turn freely in clockwise direction and lock when turned in counterclockwise direction.

k. Inspect field coil for damage, bum marks, or wear shown on pole pieces. Replace starter if damaged.

323.5. Assembly.

a. Place two thrust washers (30, FIGURE 3-47) on commutator end of armature (18) and one thrust washer (7) on other end of armature.

NOTE

When installing rotating or sliding parts, ensure they move smoothly.

NOTE

Thrust washers (30) consist of two washers, one 0.008 in. (0.2 mm) and the other 0.039 in. (1.0 mm) thick. Thrust washer (31) is 0.009 in. (0.25 mm) thick. Thrust on armature is normally 0.004 in. (0.1 mm) to 0.019 in. (0.5 mm). If it exceeds 0.019 in. (0.5 mm) an additional thrust washer is required. Armature thrust should not be removed completely.

b. Install center bracket (28) and inside metal (29) on armature shaft.

c. Install clutch (27) pinion stopper (26) and washer (25) on armature shaft.

d. Slide armature assembly into field coil (24).

e. Install shift fork (20) as noted during disassembly.

f. Install gear case (23).

g. Secure shift fork (20) with stud (19) new lockwasher (17) and nut (16).

h. Install dust covers (21 and 22).

i.   Position brush holder (15). Use care not to damage negative brushes (14).

j. Lift springs (13) on negative brushes (14) and let brushes rest on commutator. Release springs to apply tension on brushes.

k. Lift springs (13) and insert positive brushes (12) into brush holders. Release springs to apply tension on brushes.

l.   Position housing (11) on field coil and armature. Align benchmarks made during disassembly. Secure with two bolts (9) and washers (10).

m.   Install two screws (6).

n. Position magnetic switch (5) in gear case (23) and secure with two bolts (4).

o. Connect field coil connecting wire to terminal on magnetic switch (5).

3-23.6. Installation.

a. Position starter on flywheel housing engaging clutch (27, FIGURE 3-47) gear teeth with flywheel ring gear. Secure starter to flywheel housing with two nuts (1) new lockwashers (2) and washers (3).

b. Connect electrical leads to starter as tagged during removal.

c. Start engine and check for proper operation.

3-24. ALTERNATOR.

3-24.1. Battery Charging Alternator Test (Installed)

a. Check for battery voltage on alternator between terminals "BAT" and "R" and ground with the master switch (S1) in the Prime Run Position (Note Voltage).

b. Start Gen Set operator and recheck voltage on alternator terminal Pos Bat and ground for 28 + 2 VDC.

3-24.2. Removal.

a. Tag and disconnect electrical leads from alternator.

b. Loosen screw (1, FIGURE 349) and two nuts (4). Pivot alternator to relieve tension on fan belt and remove fan belt from pulley (13).

c. Remove screw (1) and washer (2) securing alternator to bracket (7).

d. Support weight of alternator and remove two bolts (3) washers (6), lockwashers (5) and nuts (4); remove alternator from lower mounting bracket. Discard lockwashers (5).

e. If necessary, remove screw and washer securing bracket (7) to water pump.

3-24.3. Disassembly.

a. Benchmark front cover (10, FIGURE 3-49) stator (25) and rear cover (24) to aid in aligning parts during assembly.

```
┌─────────────────────┐
│      CAUTION        │
└─────────────────────┘
```

Ensure stator (25) and rear cover (24) remain together to prevent damage to wires connecting stator to rectifier (26).

b. Remove three thru-screws (8 and 9). Remove front cover (10) with rotor (11) from stator (25) and rear cover (24).

c. Clamp rotor (11) in a soft-jawed vise.

d. Remove nut/lockwasher (12) securing pulley (13) to rotor shaft and remove pulley (13).

e. Remove spacer (14) and front cover (10) from rotor.

f. Remove three screws (15) securing retainer (16) to front cover. Remove retainer (16) and bearing (17) from front cover.

NOTE

If bearing (18) is removed from shaft, bearing must be replaced.

g. Remove bearing (18) from rotor shaft.

h. Remove screw (19) nuts (20) nuts (21 and 22) washers (30 and 31) and insulator (23).

i. Carefully remove rear cover (24) from stator (25).

j. Disconnect rectifier (26) wiring from stator using a soldering gun. Melt the solder (SN60WRAP2) quickly and use long-nosed pliers or equivalent to allow for heat dissipation.

k. Disconnect rectifier (26) from brush assembly (28) using a soldering gun. Melt the solder quickly and use long nosed pliers or equivalent to allow for heat dissipation.

l. Remove bolts (27) from brush assembly (28).

J. Disconnect brush assembly (28) from regulator (29) using a soldering gun. Melt the solder quickly and use long nosed pliers or equivalent to allow for heat dissipation.

3-24.4. Inspection/Testing.

a. Inspect pulley for cracks and wear.

b. Inspect fan for wear and damaged vanes.

c. Inspect brushes for wear. Normal length is 0.630 in. (16 mm), wear limit is 0.276 in. (7 mm). If below wear limit, replace brush assembly.

d. Check brush spring tension. Ensure they hold brush tight against slip ring. If weak, replace brush assembly.

e. Inspect/check rotor assembly as follows:

(1) Measure OD of slip rings. Standard OD is 1.245 in. (31.6 mm) and useable limit is 1.205 in. (30.6 mm). If OD is less than limit, replace rotor.

(2) Check slip ring surface for roughness, oil stain, etc. When surface is rough, use abrasive paper (#400) to smooth roughness. If dirty or oil stained, wipe clean with alcohol soaked cloth.

(3) Set multimeter for ohms and connect to each slip, ring and check for continuity. Standard rotor coil resistance is 12.72 ohms. If there is no continuity, rotor coil is open and rotor must be replaced.

(4) Connect multimeter from one slip ring to shaft. If there is continuity, rotor coil is shorted to ground and rotor must be replaced.

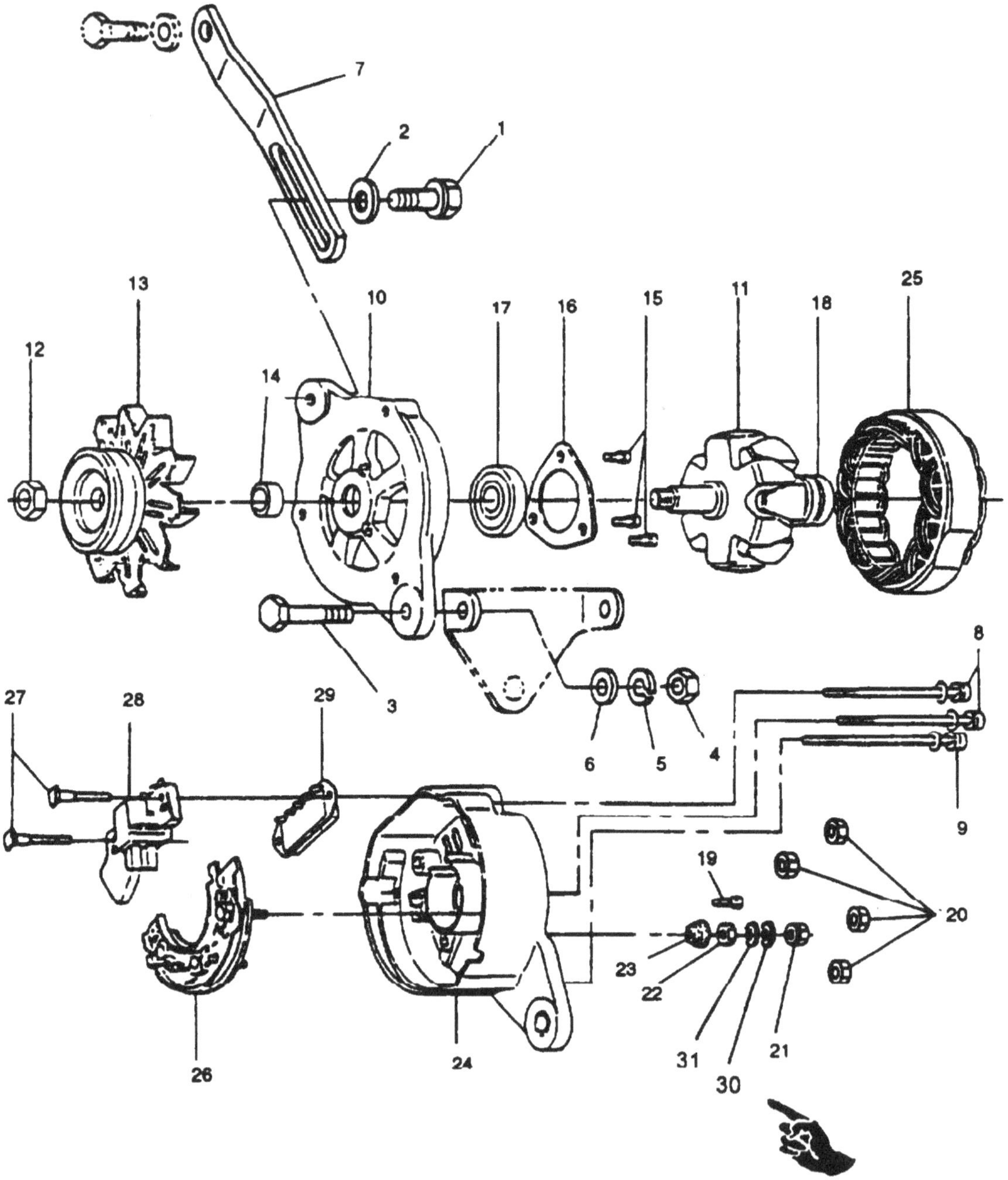

FIGURE 3-49. Alternator Assembly

f. Check stator as follows:

(1) Connect multimeter to each set of stator leads and check for continuity. Standard stator coil resistance is 0.402 ohms. If there is no continuity, stator coil is open and stator must be replaced.

(2) Connect multimeter from stator lead to stator frame. If there is continuity, stator core is shorted to ground and stator must be replaced.

g. Check rectifier as follows:

(1) Check positive side diodes by connecting multimeter positive test probe to each stator terminal and nega-tive test probe to BAT terminal. There should be no continuity. Reverse test probes. There should be con-tinuity.

(2) Check negative side diodes by connecting multimeter positive test probe to each stator terminal and neg-ative test probe to terminal E. There should be continuity. Reverse test probes. There should be no conti-nuity.

(3) If any diode is found to be bad, replace rectifier.

3-24.5. Assembly.

a. When soldering leads, use solder (SN60WRAP2), a 572 to 662°F (300 to 350°C) soldering iron (gun), and complete each solder joint within 5 seconds. Use long nosed pliers or equivalent to allow for heat dissipation.

b. Connect regulator (29, FIGURE 3-49) to brush assembly (28) and solder all connections.

c. Press bolts (27) into brush assembly (28) and regulator (29). Pressure required is 221 lbs (100 kg).

d. Connect rectifier (26) to brush assembly (28) and solder all connections.

e. Connect stator leads to rectifier (26) and solder all connections.

f. Mate stator (25) with end cover (24) aligning benchmarks.

g. Install four nuts (20), insulator (23). two nuts (21 and 22), and screw (19). Tighten hardware to 32.5 in-lbs (3.7 Nm).

h. To keep stator and end cover aligned, place a steel rod through two of the thru-screw holes until ready to insert thru-screws.

i. Install new bearing (18) on rotor (11) shaft.

j. Install bearing (17) in front cover (10) . Install retainer (16) and secure with three screws (15). Tighten screws to 32.5 in-lbs (3.7 Nm).

k. Install front cover (10) and spacer (14) on rotor (11) shaft.

l. Clamp rotor (11) in a soft-jawed vise.

m. Install pulley (13) on rotor shaft and secure with nut/lockwasher (12). Tighten nut to 38 ft-lbs (51.5 Nm).

NOTE

Use small pin through hole in rear cover to hold brush depressed in brush holder to allow slip ring to pass through.

n. Position rotor (11) with front cover (10) attached into stator and rear cover. Align benchmarks, remove two steel rods, and secure with three thru-screws (8 and 9). Tighten screws to 32.5 in-lbs (3.7 Nm).

3-24.6. Installation.

a. If removed, position bracket (7, FIGURE 3-49) on water pump mounting and secure with bolt and washer. Tighten bolt to 30 ft-lbs (40.6 Nm).

b. Position alternator on tower mounting bracket and secure with two bolts (3), washers (6). new lockwashers (5), and nuts (4). Do not tighten bolts.

c. Attach alternator to bracket (7) with screw (1) end washer (2).

d. Pivot alternator and route fan belt over pulley (13). Adjust belt tension, refer to end item maintenance manual.

e. Connect electrical leads to alternator as tagged during removal.

f. Operate engine and check for proper operation of electrical system.

SECTION IX. ENGINE BLOCK MAINTENANCE

3-25. <u>FRONT GEAR COVER.</u>

3-25.1. <u>Removal.</u>

 a. Drain engine lubrication system. Refer to end item maintenance manual.

 b. Remove fan and fan belt, refer to end item maintenance manual.

 c. Remove four screws (1, FIGURE 3-50) cover (2) and preformed packing (3). Ensure "ZZ' is still visible.

 d. Rotate crankshaft pulley (5) until timing mark is lined up with dial pointer (6), refer to FIGURE 3-6.

NOTE

Ensure engine is at TDC.

 e. Remove bolt (4, FIGURE 3-50) securing crankshaft pulley (5) to crankshaft and remove pulley.

 f. Remove two screws (7).

 g. Support front gear cover (9) and remove thirteen screws (8) attaching cover to engine block; remove cover (9) and preformed pecking (10). Discard preformed packing.

 h. Remove dial pointer (6).

 i. Remove and discard two preformed packings (11) from front gear cover.
Use care not to damage oil seal fitting surfaces.

$$\boxed{\textbf{CAUTION}}$$

 i.Remove crankshaft front oil seal (12) from front gear cover using a plastic hammer and a screwdriver to tap around oil seal to free it.

3-25.2. <u>Inspection.</u>

 a. Inspect crankshaft front oil seal (12, FIGURE 3-50) for excessive wear, damage, or signs of leakage.

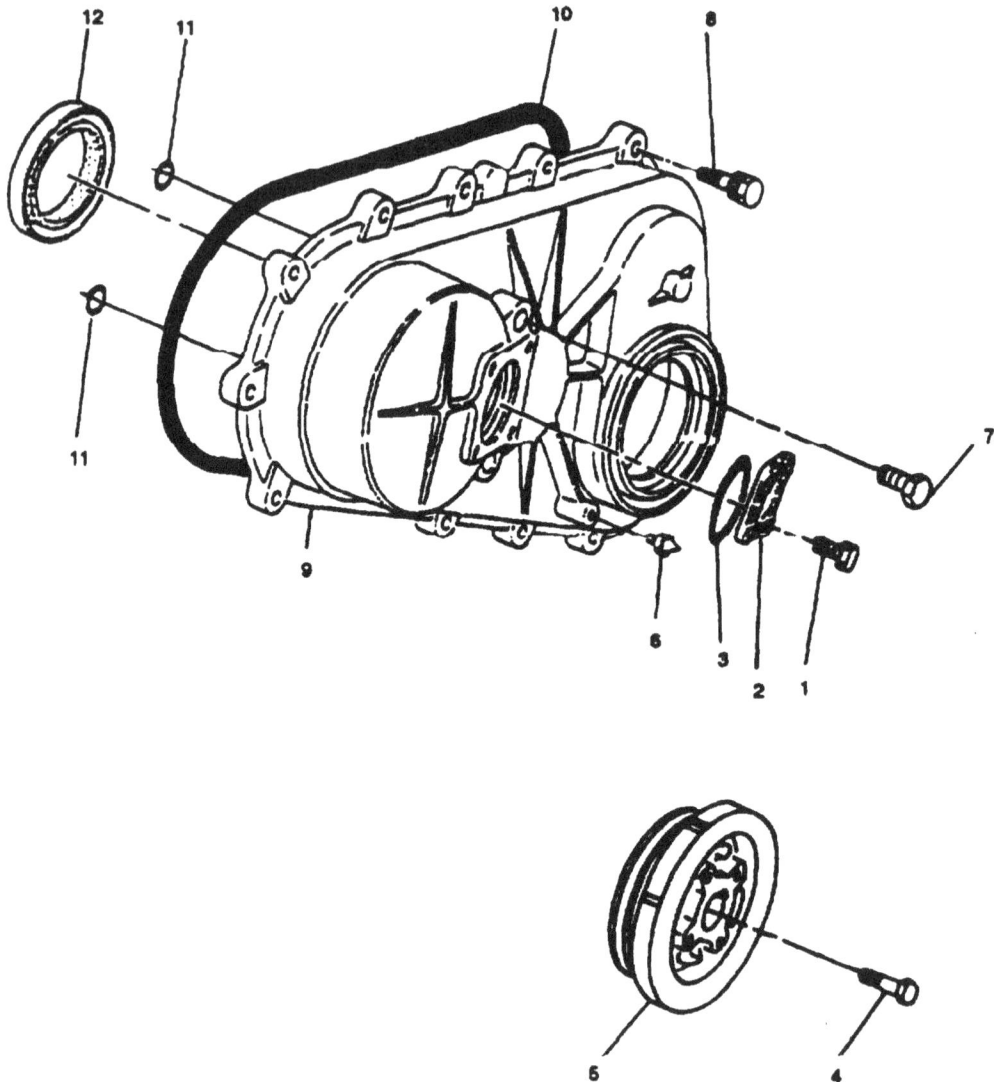

FIGURE 3-50. Front Gear Cover

b. Inspect crankshaft pulley (5) for cracks and wear.

c. Inspect front gear cover (9) for cracks or other physical damage.

3-25.3. Installation.

   a. Install new crankshaft front oil seal (12, FIGURE 3-50) in front gear cover (9) using oil seal installer (5-6522-9613-0)

   b. Install two new preformed packings (11) in front gear cover.

   c. Install dial pointer (6).

   d. Insert new preformed packing (10) firmly into groove on front gear cover.

   e. Position front gear cover (9) on engine block and secure with thirteen screws (8). Tighten each bolt to 168 ftlbs (19.9 Nm) in the sequence shown in FIGURE 3-51.

   f. Install two screws (7).

NOTE

Verify engine is at TDC by viewing "ZZ" mark on timing gear visible through front cover access panel.

   g. Install crankshaft pulley (5, FIGURE 3-50) on crankshaft, aligning mark on pulley with pointer and secure with bolt (4). Tighten bolt to 137 ft-lbs (166.0 Nm).

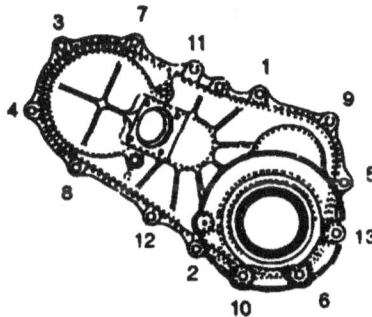

FIGURE 3-51. Front Gear Cover Screw lightening Sequence

h. Install new preformed gasket (3) and cover (2). Secure cover (2) with four screws (1).

i. Install fan belt and fan, refer to end item maintenance manual.

j.Service engine lubrication system, refer to end item maintenance manual. 3-26._

ROCKER COVER AND ARMS. 3-26.1. Removal.

a. Remove PCV assembly as required to access rocker cover, refer to paragraph 3-20.1.

NOTE

Rocker arm shaft (18) positioning is done by a cutaway machined on shaft end. Install retaining plates (13) and shaft (18) so that cutaway faces mating bolt stem.

NOTE

Rocker arms (12, 14, 16, and 17) each have a different shape. Be sure to install them in their correct location (refer to Figures 3-52 and 3-53) as viewed from front.

b. Remove two bolts (1, FIGURE 3-52) washers (2), and gaskets (3) securing rocker cover (5); remove rocker cover (5) and preformed packing (6). Discard two gaskets (3) and preformed packing (6).

c. Loosen eight nuts (19) and loosen adjusting screws (20) until the adjusting screws (20) just touch pushrods.

d. Remove valve rocker oil feed pipe, refer to paragraph 3-16.1.
Failure to loosen retainer plate bolts (7) a little at a time in the proper sequence, will adversely affect rocker arm shaft (18).

CAUTION

e. Loosen eight bolts (7) a little at a time in the sequence shown in FIGURE 3-53.

f. After all bolts have been loosened, remove eight bolts (7, FIGURE 3-52) and washers (8). Remove rocker arm assembly.

3-26.2. Disassembly.

a. Remove two retaining rings (9, FIGURE 3-52) and washers (10) from each end of shaft (18).

b. Remove eight rocker arms (12, 14, 16, and 17) four retaining plates (13) and three springs (15) from shaft (18).

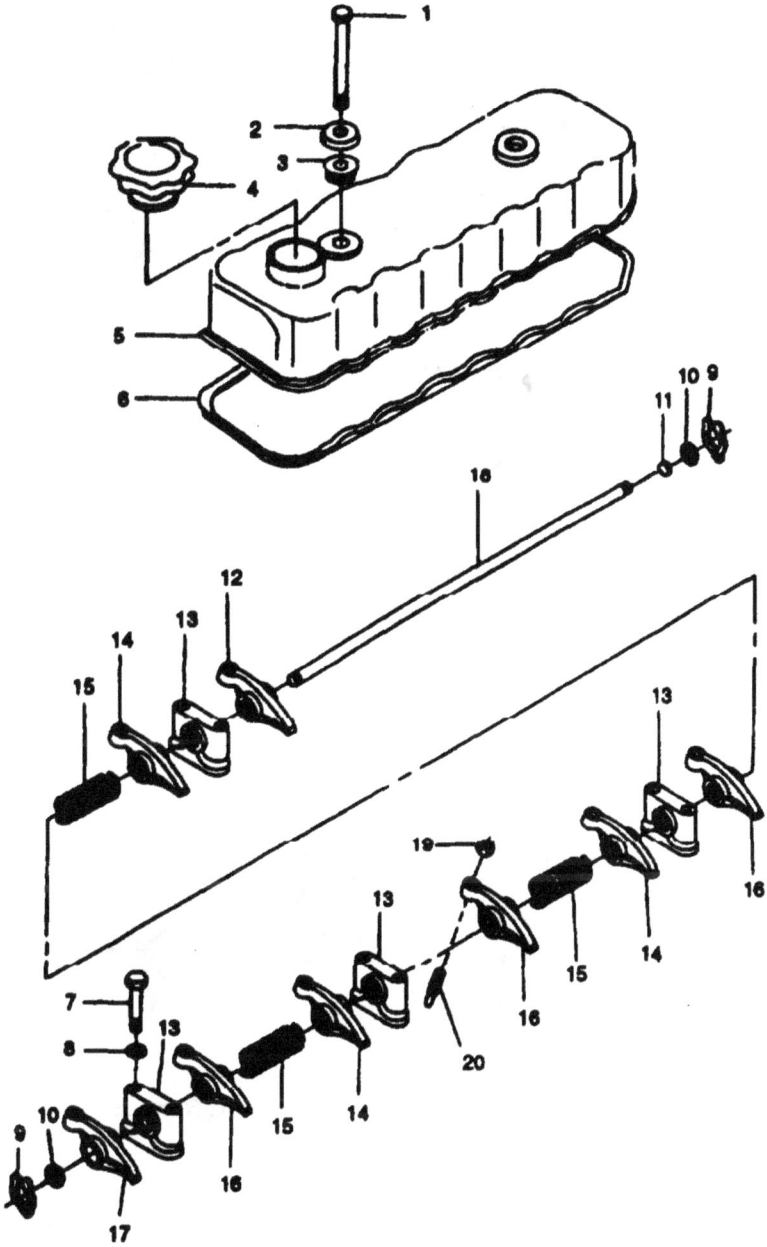

FIGURE 3-52. Rocker Covet and Arms

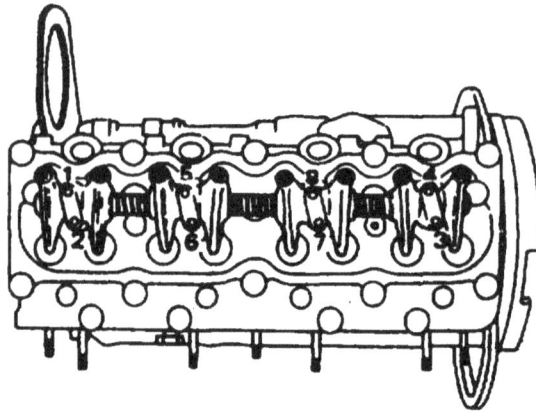

FIGURE 3-53. Rocker Arm Retaining Plate Bolt Loosening Sequence

c. Remove nut (19) and adjusting screw (20) from each rocker arm.

d. Remove dust cap (11) from each end of shaft (18).

3-26.3. Inspect and Measure.

a. Check rocker arm shaft (18, FIGURE 3-52) run-out as follows:

(1) Place rocker arm shaft on V-blocks, refer to FIGURE 3-54.

FIGURE 3-54. Rocker Arm Shaft Runout

(2) Use a dial indicator to measure shaft central portion run-out. Limit on run-out is 0.024 in. (0.6 mm). If run-out exceeds limit, replace shaft. If run-out is very slight, an arbor press may be used to correct shaft. Shaft must be cold.

b. Use a micrometer and measure rocker arm shaft outside diameter, refer to FIGURE 3-55. The standard diam-eter is 0.747 to 0.748 in. (18.98 to 19.0 mm), with a limit of 0.744 in. (18.90 mm). If measured value is less than limit, replace rocker arm shaft (18, FIGURE 3-52).

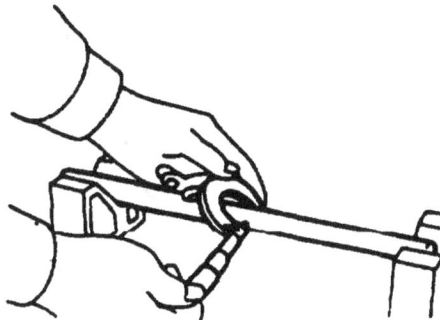

FIGURE 3-55. Rocker Arm Shaft Outside Diameter

c. Use either a vernier caliper or a dial indicator and measure rocker arm (12, 14, 16, and 17, FIGURE 3-52) bushing inside diameter. The standard diameter is 0.749 to 0.750 in. (19.036 to 19.060 mm), with a limit of 0.752 in. (19.10 mm). If measured value exceeds limit, replace rocker arm.

d. Clearance between rocker arm shaft 00 and rocker arm bushing ID standard value is 0.0016 in. (0.04 mm). If clearance exceeds 0.008 in. (0.20 mm), replace either rocker arm shaft (18) or rocker arm (12, 14, 16, and 17).

```
WARNING
```

Compressed air used for cleaning can create airborne particles that may
enter the eyes. Pressure will not exceed 30 psig (207 kPa). Eye protection
required.

e. Check that rocker arm oil ports (two) are free of obstructions. refer to FIGURE 3-56. Use compressed air if
necessary to clean oil ports.

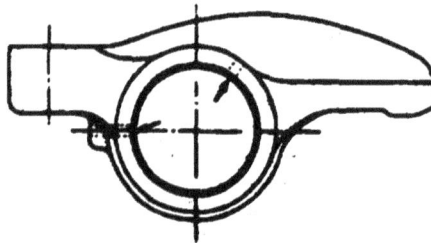

FIGURE 3-56. Rocker Arm Oil Ports, Location of

f. Inspect rocker arm valve stem contact surface for step wear and storing. If contact services have light step
wear or scoring, they may be honed with oil stone. If step wear or scoring is severe, rocker arm must be
replaced, refer to FIGURE 3-57.

3-26.4. Assembly.

```
WARNING
```

Compressed air used for cleaning can create airborne particles that may
enter the eyes. Pressure will not exceed 30 psig (207 kPa). Eye protec-

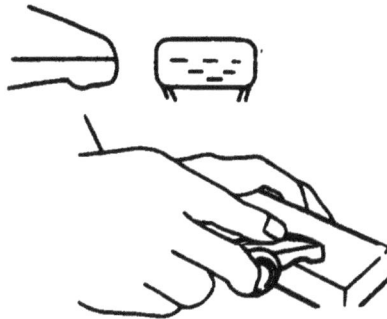

FIGURE 3-57. Rocker Arm Correction

a. Use compressed air to thoroughly clean rocker arm shaft (18, FIGURE 3-52) and rocker arms (12, 14, 16, and 17) oil ports.

b. Coat rocker arm shaft and rocker arm bushings with engine lubricating oil (MIL-L-2184).

NOTE

Rocker arm shaft (18) positioning is done by a cutaway machined on shaft end. Install retaining plates (13) and shaft (18) so that cutaway faces mating bolt stem.

NOTE

Rocker arms (12, 14, 16, and 17) each have a different shape. Be sure to install them in their correct location (refer to FIGURES 3-52 and 3-53) as viewed from front.

c. Install dust plugs (11) in each end of shaft (18).

d. Install adjusting screw (20) and nut (19) in each rocker arm so they are in a loosened condition.

NOTE

Back end of shaft (18) is the end with the oil supply hole.

e . Install retaining ring (9) and washer (10) on one end of shaft (18). This will be back end of shaft (in relation to engine).

f. Installone rocker arm (12), three rocker arms (14) three rocker arms (16), one rocker arm (17), four retaining plates (13) and three springs (15) following the sequence shown in FIGURES 3-52 and 3-53. Start with rocker arm (12) and finish with rocker arm (17) (front of engine). Retaining plates are marked with an 'F and this side must face the front of engine.

g. Install washer (10) and retaining ring (9) on end of shaft (18)

3-26.5. Installation.

a. Position rocker arm assembly on cylinder head and secure four retaining plates (13, FIGURE 3-52) with two bolts (7) and washers (8) each. Tighten bolts to 168 in-lbs (19.0 Nm) in sequence shown in FIGURE 3-58.

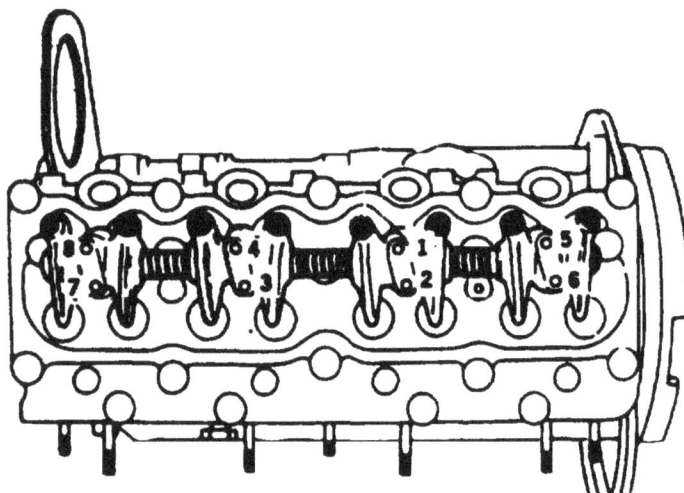

FIGURE 3-58. Rocker Arm Retaining Plate Bolt Tightening Sequence

b. Install PCV assembly, refer to paragraph 3-20.3.

c. Install valve rocker oil feed pipe, refer to paragraph 3-16.2.

d. Perform adjustment procedures, refer to paragraph 3-26.6.

e. Position new preformed packing (6) and rocker cover (5) on cylinder head and secure with two bolts (1), washers (2) and new gaskets (3). Tighten bolts to 114 in-lbs (13.0 Nm)

3-26.6. Adjustment.

a. Remove rocker cover (5, FIGURE 3-52) if not already removed, refer to paragraph 3-26.1.

b. Ensure all eight retaining plate bolts (7) are tight. If not, tighten in accordance with paragraph 3-26.5.

c. Rotate crankshaft until crankshaft pulley TDC line is aligned with timing pointer. This will bring piston in No. 1 or No. 4 cylinder to TDC on compression stroke.

d. Check for play in No. 1 intake and exhaust valve push rods. If push rods have play, No. 1 piston is at TDC on compression stroke. If push rods are depressed, No. 4 piston is at TDC on compression stroke.

e. Adjust valve clearance for rocker arms shown in View A of FIGURE 3-59 as follows:

   (1) Loosen nut (19, FIGURE 3-52) and adjusting screw (20).

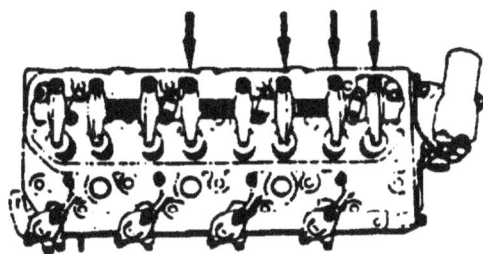

VIEW A

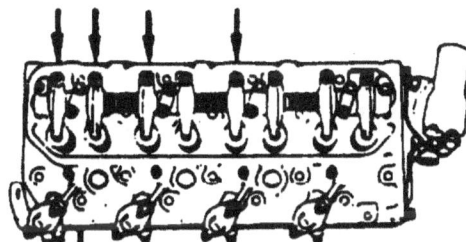

VIEW B

FIGURE 3-59. Valve Clearance Adjustment Sequence

(2) Insert a 0.018 in. (0.45 mm) feeler gage between rocker arm and end of valve stem.

(3) Turn adjusting screw (20) until a slight drag is felt on feeler gage.

(4) Tighten nut (19).

f. Rotate crankshaft 360 degrees until crankshaft pulley TDC fine is aligned with timing pointer.

g. Adjust valve clearance for rocker arms shown in View B of FIGURE 3-59 as described in step e.

h. Install rocker cover (5, FIGURE 3-52) refer to paragraph 3-26.5. 3-27.

CYLINDER HEAD ASSEMBLY. 3-27.1. Removal.
Do not drain coolant until coolant temperature is below operating tem-perature. Scalding can

| **WARNING** |

occur from failure to observe this warning.

a. Drain engine coolant and lubrication systems, refer to end item maintenance manual.

b. Remove thermostat housing, refer to paragraph 3-6.1.

c. Remove fuel injector nozzles and piping, refer to paragraph 3-10.1.

d. Remove glow plugs, refer to paragraph 3-22.1.

e. Remove intake and exhaust manifolds, refer to paragraph 3-21.1.

f. Remove rocker cover and rocker arms, refer to paragraph 3-28.1.

g. Remove eight pushrods (1, FIGURE 3-60).

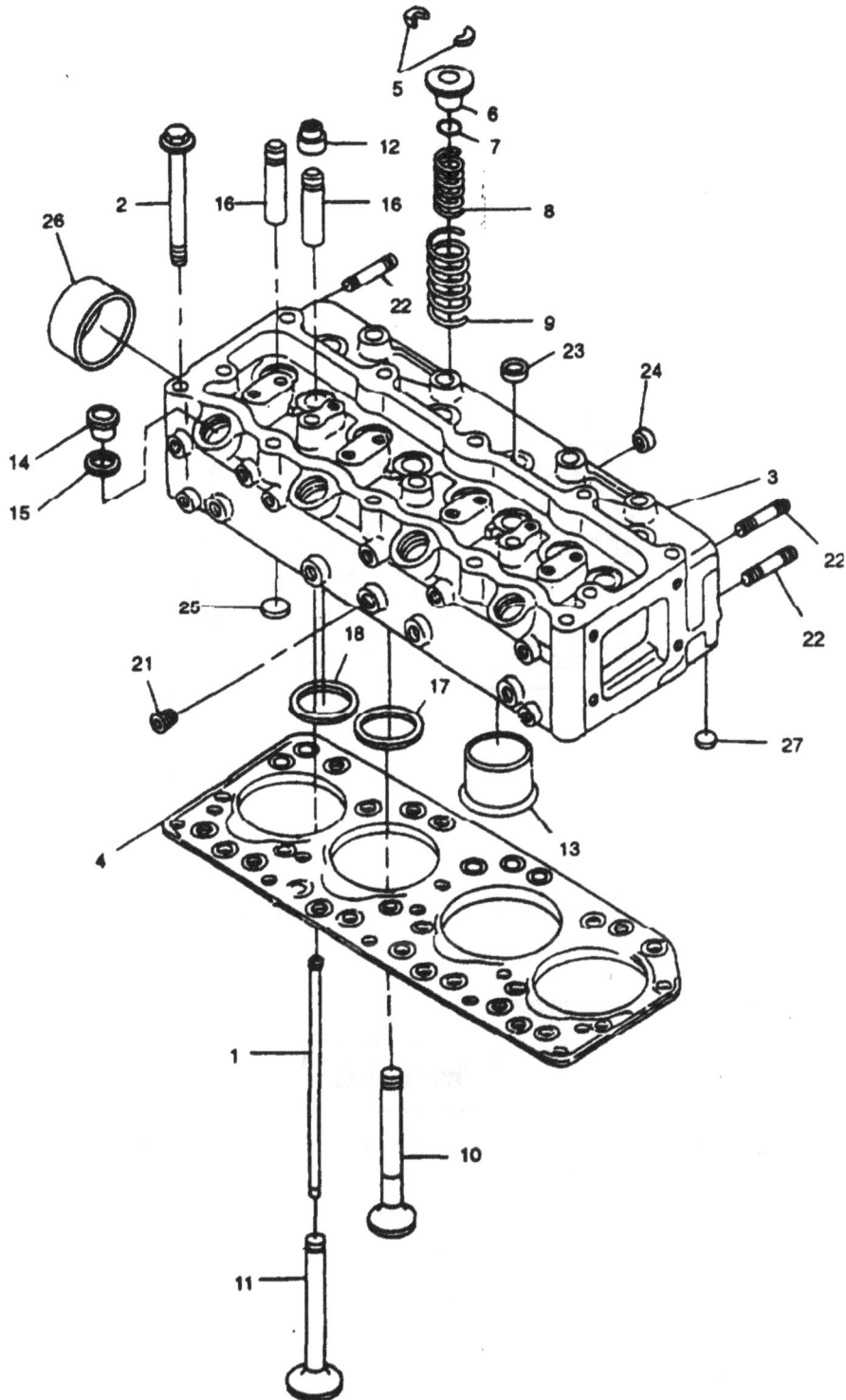

FIGURE 3-60. Cylinder Head Assembly

```
┌──────────────┐
│   CAUTION    │
└──────────────┘
```

Failure to loosen cylinder head bolts (2) a little at a time and in the proper sequence will adversely affect cylinder head (3) lower surface.

h. Loosen nineteen bolts (2) a file at a time in the sequence shown in FIGURE 3-61.

i.   After all nineteen bolts (2, FIGURE 3-60) have been loosened, remove them and lift cylinder head assembly

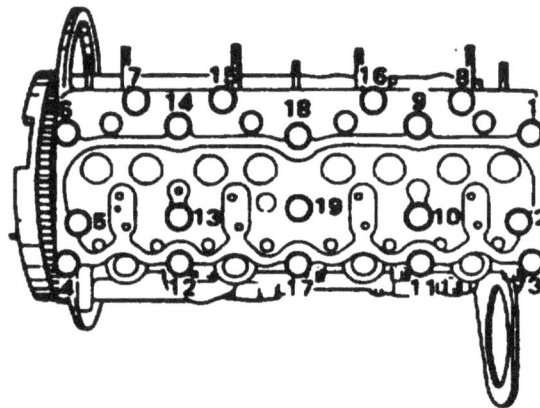

FIGURE 3-61. Cylinder Head Bolt Loosening Sequence

3-27.2. Disassembly.

a. Place cylinder head on a flat wooden surface.

```
┌──────────────┐
│   WARNING    │
└──────────────┘
```

To avoid personal injury, do not stand in front of valve springs while

```
CAUTION
```

Do not allow valves (10 and 11, FIGURE 3-60) to fall from bottom of cylinder head.

b. Use spring compressor to compress springs (8 and 9) and remove spring lock (5).

c. Release spring compressor and remove seat assembly (6), sealing ring (7). and springs (8 and 9) from valve stem.

d. Repeat steps b and c for remaining seven valves.

NOTE

If valves are to be reused, tag them with cylinder number from which they were removed. If they are to be replaced, valve guides must also be replaced.

e. Remove eight valves (10 and 11).

f. Remove eight valve seals (12) from valve guides (16).

g. Remove hot plugs (13) as follows:

(1) Refer to FIGURE 3-62 and insert a 0.12 to 0.20 in. (3.0 to 5.0 mm) diameter brass bar into fuel injector fifing until it makes contact with hot plug (13, FIGURE 3-60).

(2) Lightly tap bar with a hammer to drive hot plug free.

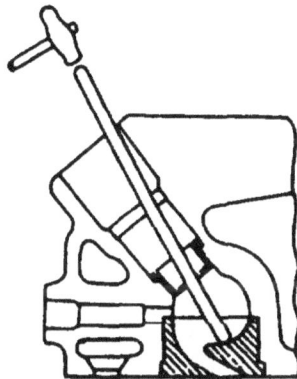

FIGURE 3-62. Hot Plug Removal

h. After hot plug has been removed, use hammer and same brass bar through hot plug hole to lightly tap lower side of heat shield (14) and drive it free. Remove heat shield (14) and washer (15).

i. Remove valve guides (16) using a hammer and valve guide replacer (NU-7634) to drive out guides from lower face of cylinder heed.

j. Remove valve seat inserts (17 and 18) as follows:

(1) Using MIG welder (refer to TM 9-239) weld four spots evenly spaced to inside circumference of valve seat insert. These will be used to pry insert free.

(2) Allow it to cool for a few minutes which will cause contraction and make removal easier.

> **CAUTION**

Use care not to damage cylinder head.

(3) Use a screwdriver and pry valve seat insert free.

(4) Carefully remove carbon and other foreign material from cylinder head insert bore.

3-27.3. Inspect and Measure.

a. Check cylinder head lower face warpage as follows:

(1) Use a straight edge and feeler gage to measure the four sides and two diagonals of cylinder head lower face, refer to FIGURE 3-63.

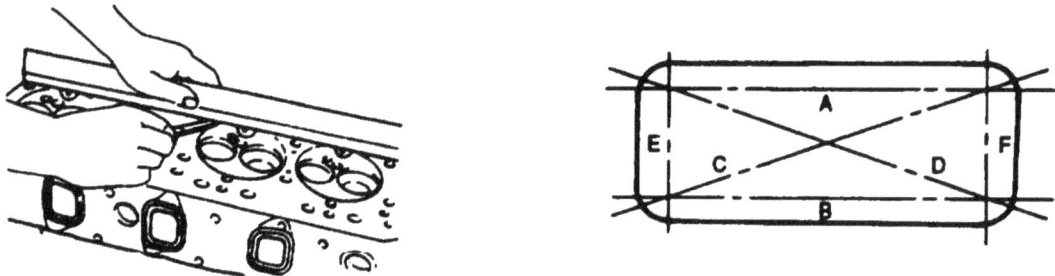

FIGURE 3-63. Cylinder Head Inspection

(2) Standard warpage should be 0.002 in. (0.05 mm) or less.

(3) Regrind cylinder head lower face if measurement values are greater than warpage limit of 0.008 in. (0.2 mm), but less than maximum grinding limit of 0.012 in. (0.30 mm)

(4) If warpage exceeds maximum grinding limit, cylinder head must be replaced.

(5) Standard cylinder height as measured on fuel injector side of head is 3.620 to 3.624 in. (91.95 to 92.05 mm), with a limit of 3.612 in. (91.75 mm), refer to FIGURE 3-64.

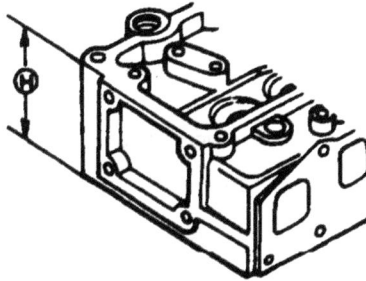

FIGURE 3-64. Cylinder Head Height (Reference)

(6) If cylinder head lower face is reground, valve depression must be checked in accordance with step h.

b. Check hot plug depression as follows:

(1) Clean cylinder head lower face, using care not to damage hot plug surfaces, refer to FIGURE 3-65.

(2) Use a straight edge and feeler gage to measure hot plug depression in a straight line from No. 1 to No. 4 hot plug.

(3) If measured value exceeds 0.0008 in. (0.02 mm), replace hot plugs.

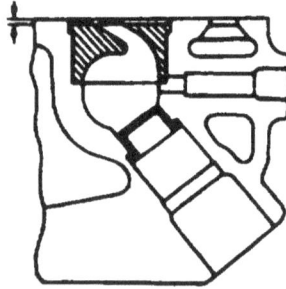

FIGURE 3-65. Hot Plug Depression

c. Inspect combustion chamber as follows:

    (1) Remove carbon adhering to inside of combustion chamber. Use care not to damage hot plug fitting positions.

    (2) Inspect combustion chamber, hot plug hole, and hot plug machined faces for cracks and other damage.

    (3) If cracks or other damage are found, cylinder heed must be replaced.

NOTE

Be absolutely certain there are no scratches or protrusionson combustion chamber surfaces which will be in contact with hot plug after it is installed. These flaws will prevent hot plug from seating correctly.

d. Inspect hot plugs for excessive wear and other damage. Replace hot plugs if any of these conditions are found.

e. Check valve stem and valve guide clearance as follows.

    (1) With valve and valve guide in cylinder head, set a dial indicator at zero on valve stem, refer to FIGURE 3-66.

    (2) Move valve heed from side to side and reed dial indicator. Note highest reading.

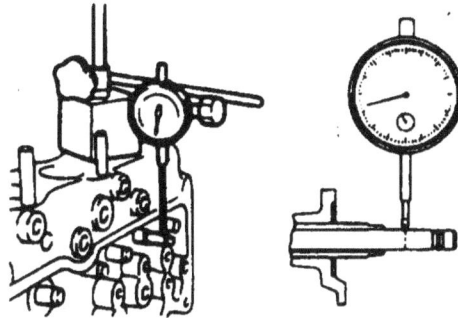

FIGURE 3-66. Valve Stem and Guide Clearance

(3) If measured values exceed specified limit, valve and valve guide must be replaced as a set.

(4) Intake valve stem clearance standard is 0.0015 to 0.0027 in. (0.039 to 0.069 mm), with a limit of 0.008 in. (0.20 mm). Exhaust valve stem clearance standard is 0.0025 to 0.0037 in. (0.084 to 0.093 mm), with a limit of 0.0098 in. (0.25 mm).

f. Check valve stem outside diameter as follows:

(1) Using a micrometer, measure valve stem OD at three points, refer to FIGURE 3-67.

FIGURE 3-67. Valve Stem Outside Diameter

(2) If measured value is less than limit, replace valve and valve stem as a set.

(3) Intake valve stem standard OD is 0.3129 to 0.3134 in. (7.949 to 7.961 mm), with a limit of 0.3102 in. (7.88 mm). Exhaust valve stem standard OD is 0.3118 to 0.3124 in. (7.921 to 7.936 mm), with a limit of 0.3090 in. (7.85 mm)

g. Check valve thickness, refer to FIGURE 3-68. Standard measurement is 0.05 in. (1.3 mm), with a limit of 0.04 in. (1.0 mm). If measured value is less than limit, replace valve and valve guide as a set.

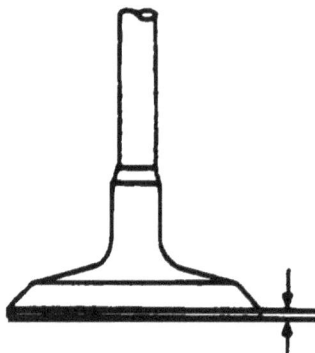

FIGURE 3-68. Checking Valve Thickness

h. Check valve depression as follows:

(1) Install valve in cylinder head, refer to FIGURE 3-69.

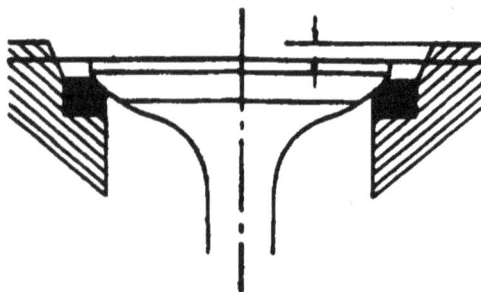

FIGURE 3-69. Valve Depression

(2) Use a depth gage or straight edge with steel rule to measure valve depression from cylinder head lower surface.

(3) The standard depression is 0.028 in. (0.7 mm), with a limit of 0.11 in. (2.7 mm). If depression exceeds limit, replace valve seat insert.

i.   Check valve contact width as follows:

(1) Check valve contact faces for roughness and unevenness. Smooth contact surfaces as necessary.

(2) Using blue bearing dye (HISPOTBLUE 107), coat valve seat insert. Install valve and rotate valve to pickup chalk on valve.

(3) Measure valve contact width. Refer to FIGURE 3-70. Standard width is 0.05 to 0.06 in. (1.2 to 1.5 mm), with a limit of 0.08 in. (2.0 mm).

(4) If measured value exceeds limit, replace valve seat insert.

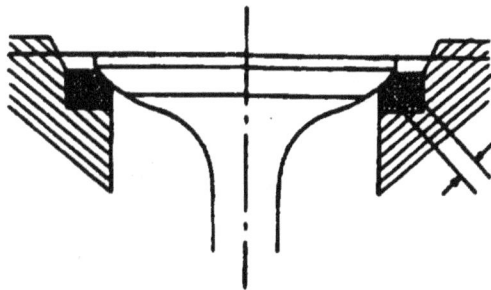

FIGURE 3-70. Checking Valve Contact Width

j.   Check valve spring free length using a vernier caliper, refer to FIGURE 3-71. Standard height of inner spring is 1.89 in. (47.9 mm), with a limit of 1.83 in. (46.5 mm). Standard height of outer spring is 1.86 in. (47.3 mm), with a limit of 1.80 in. (45.8 mm). If measured value is less than limit, replace spring.

k. Check valve spring inclination using a surface plate and a square. If inclination measured exceeds 0.04 in.

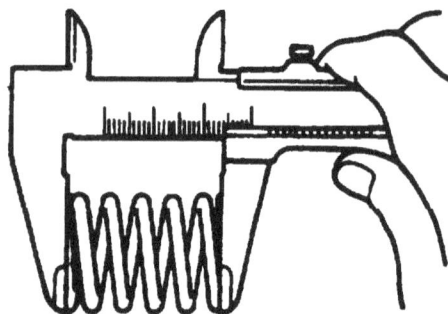

FIGURE 3-71. Valve Spring Free! Height

I . Check valve spring tension as follows:

(1) Install spring in spring tester, refer to FIGURE 3-72. Compress innerspring to 1.46 in. (37.0 mm) or outer spring to 1.54 in. (39.0 mm).

(2) Standard pressure tocompress inner spring is 12.21 to 13.78 lbs (5.55 to 6.25 kg), with a limit of 11.03 lbs (5.0 kg). Standardpressure to compress outer spring is 43.33 to 48.84 lbs (19.65 to 22.15 kg). with a limit of 39.69 lbs (18.0 kg)

(3) If pressure required tocompress spring is less than limit, replace spring.

FIGURE 3-72. Valve Spring Tension

m. Check pushrod curvature as follows:

(1) Lay pushrod on a surface plate, refer to FIGURE 3-73.

NOTE

If surface plate is not available, use a flat surface.

(2) Roll pushrod along surface plate checking for curvature with a thickness gage.

(3) If measured value exceeds 0.012 in. (0.3 mm), replace pushrod.

(4) Inspect both ends of pushrod for excessive wear or damage. If excessive wear or damage is found, replace pushrod.

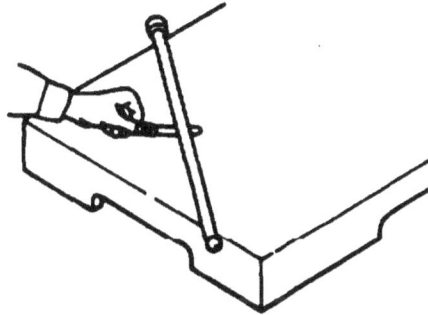

FIGURE 3-73. Push Rod Curvature

n. Inspect studs (22, FIGURE 3-60) for stripped threads or other damage.

o. Inspect freeze plugs (23, 24, 25, and 26) and dust plug (27) for cracks and other damage.

3-27.4. Repair.

a. Correct valve seat insert as follows:

(1) Remove carbon from valve seat insert surface.

(2) Use a valve cutter (15, 45, and 75 degree blades) to minimize scratches and other rough areas. This will bring contact width back to standard value, refer to FIGURE 3-74.

(3) Remove only scratches and rough areas. Do not cut away too much. Take care not to cut away unblemished areas of valve seat surface.

NOTE

Use an adjustable valve cutter pilot. Do not allow it to wobble inside valve guide.

(4) Apply lapping compound (SSL1682) to valve seat insert surface.

(5) Insert valve into valve guide.

(6) Turn valve while tapping it to fit valve seat insert.

(7) Check that valve contact width is correct.

(8) Check that valve seat insert surface is in contact with entire circumference of valve.

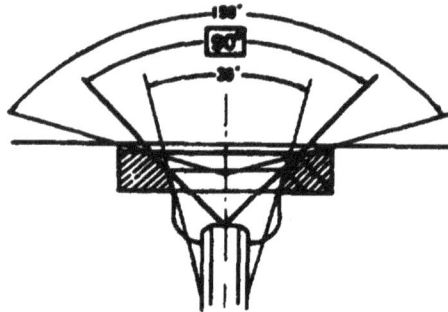

FIGURE 3-74. Valve Seat Correction

b. Replace damaged studs (22, FIGURE 3-60) freeze plugs (23, 24, 25, and 26), and duet plug (27).

3-27.5. Assembly.

    a. Install valve seat inserts (17 and 18, FIGURE 3-60) as follows:

        (1) Place valve seat insert in position and carefully place attachment (1, FIGURE 3-75), having a smaller OD than valve seat insert, on valve seat insert (2).

NOTE

Smooth side of attachment must contact insert.

CAUTION

Do not apply an excessive amount of pressure with press. Damage to valve seat insert will result.

        (2) Use an arbor press (3) to gradually apply pressure to attachment and press valve seat insert into place.

NOTE

Valve seat insert may need to be ground to obtain proper contact width.
Refer to paragraph 3-27.4.

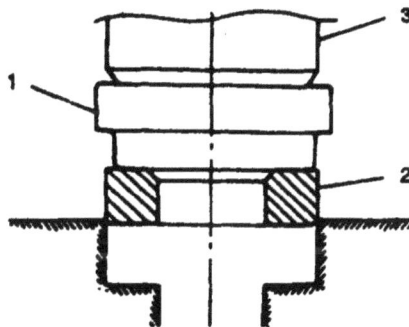

FIGURE 3-75. Valve Seat Insert Installation

b. Install valve guides (16, FIGURE 3-60) as follows:

(1) Apply engine lubricating oil (ML-L-2104) to valve guide outer circumference.

(2) Attach valve guide installing tool (NU-7634) and valve guide setting tool (320908) to valve guide.

(3) Use a hammer to drive valve guide into position from cylinder head upper face.

(4) Measure distance valve guide extends above cylinder head upper face. Distance should be approximately 0.47 in. (12.0 mm).

NOTE

If valve guide has been removed, valve and valve guide must be replaced as a set.

c. Install heat shields (14, FIGURE 3-60) as follows:

NOTE

Always install a new heat shield (14). Never reuse an old heat shield (14).

(1) Position washer (15) and heat shield (14), flange side facing up, in cylinder head through fuel injector hole.

(2) Lightly tap heat shield flange into place with a hammer and a brass bar. d .

Install used hot plugs (13) as follows:

(1) Align hot plug knock ball with cylinder head groove.

(2) Use a plastic hammer lo tap hot plug into place.

e. Install new hot plugs (13) as follows:

(1) Align hot plug knock ball with cylinder head groove and using a plastic hammer tap it temporarity into position.

(2) Place a 1.0 in. (25 mm) thick metal plate over hot plug upper surface.

CAUTION

Do not apply pressure greater than specified. Damage to cylinder head will result.

(3) Use an arbor press to exert a pressure of 9923 to 12,128 lbs (4500 to 5500 kg) on metal plate. This will drive hot plug into position.

(4) Using plastic hammer, lightly tap hot plug head to ensure it is firmly seated.

(5) Repeat steps (1) through (4) for other hot plugs.

(6) Hot plug surfaces must be perfectly flush with cylinder head lower face. Use a surface grinder to grind off any hot plug surface protrusions.

(7) After grinding, ensure hot plug surfaces are completely free of protrusions. Ensure there are no hot plug depressions.

(8) Tap each hot plug heed lightly to ensure they are firmly seated.

f. Apply a coat of engine lubricating oil (MIL-L-2104) to inner face of seal (12) and using oil seal installer, install oil seal (12) to valve guides.

g. Apply a coat of engine lubricating oil (MIL-L-2104) to valve stems and install eight valves (10 and 11) in cylinder head.

h. Install springs (8 and 9) with their fine pitched end (painted) facing down, sealing ring (7) and seat assembly (6) on valve stem.

i. Use spring compressor to compress springs and install two piece spring lock (5). i.Re-

peat steps h and i for remaining seven valves.

3-27.6. Installation.

a. Place new head gasket (4, FIGURE 3-60) on engine block with word TOP facing up.

b. Carefully place cylinder head (3) on engine block aligning engine block dowels with cylinder head dowel holes.

c. Lubricate nineteen bolts (2) with engine lubricating oil (MIL-L-2104) and install them hand tight.

d. Tighten bolts (2) in two steps following sequence shown in FIGURE 3-76.

    (1) Step 1 - New bolts - 47.0 ft-lbs

        (63.7 Nm)

      - Used bolts - 47.0 ft-lbs

        (63.7 Nm)

    (2) Step 2 - New bolts - 58 ft-lbs

        (78 Nm)

      - Used bolts - 65.0 ft-lbs

        (88.3 Nm)

e. Lubricate ends of pushrods (1, FIGURE 3-60) with engine lubricating oil (MIL-L-2104) and install eight pushrods.

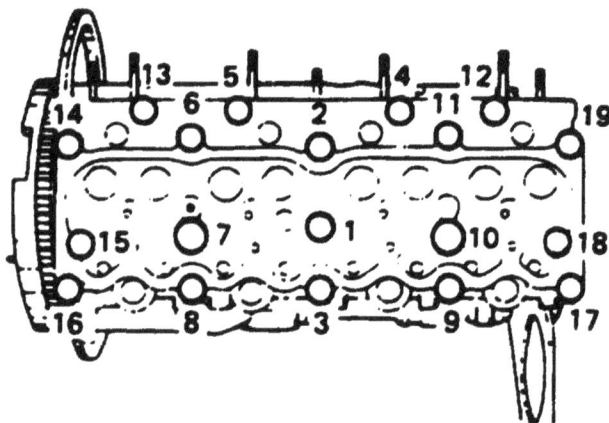

FIGURE 3-76. Cylinder Head Bolt Tightening Sequence

f. Install rocker arms, refer to paragraph 3-26.5.

g. Install intake and exhaust manifolds, refer to paragraph 3-21.3.

h. Install glow plugs, refer to paragraph 3-22.4.

i. Install fuel injector nozzles and piping, refer to paragraph 3-10.7. j. Install

thermostat housing, refer to paragraph 3-6.4.

k. Service engine coolant and lubrication systems, refer to end item maintenance manual and lubrication order. 3-27.7.

Compression Check.

a. Start engine and allow it to warm up until coolant temperature is above 176°F (80°C).

b. Shut down engine.

c. Remove glow plug from No. 1 cylinder.

d. Connect adapter (J-26999-2) and compression gage (MIL-T-2764) to No. 1 cylinder glow plug hole.

e. Turn engine over with starter and record compression gage reading. Standard compression at 200 rpm and at sea level is 441 psi (3038 kPa), with a limit of 327 psi (2254 kPa)

f. Repeat steps b through e for each cylinder.

g. If compression is less then limit, troubleshoot the engine. 3-28.

PISTONS, CONNECTING RODS, AND CRANKSHAFT 3-28.1. Removal.

h. Remove oil filter, refer to paragraph 3-14.2.

b. Remove fuel filter, refer to paragraph 3-9.1.

c. Remove starter, refer to paragraph 3-23.1.

d. Remove alternator, refer to paragraph 3-24.1.

e. Remove water pump, refer to paragraph 3-7.1.

f. Remove PCV system, refer to paragraph 3-20.1.

g. Remove oil pan, refer to paragraph 3-17.1.

h. Remove front gear cover, refer to paragraph 3-25.1.

i. Remove idler gear and shaft, refer to paragraph 3-30.1.

j. Remove fuel injection pump, refer to paragraph 3-11.1.

k. Remove cylinder heed, refer to paragraph 3-27.1.

l. Remove crankcase, refer to paragraph 3-29.1.

m. Remove oil pump refer to paragraph 3-18.1.

n. Remove carbon deposits from upper portion of cylinder wall with a scraper before removing piston and connecting rod.

o. Remove two bolts (1, FIGURE 3-77) and nuts (2) securing connecting rod bearing cap (3). Remove lower half d bearing cap (3) and connecting rod bearings (4).

NOTE

If connecting rod hearings are to be reused, tag them as to their original position to facilitate installation.

p. Move piston (5) to top of cylinder and tap it with a hammer handle or similar object, from connecting rod side, to drive it out of cylinder.

q. Repeat steps n through p for remaining pistons and connecting rode.

r. Remove camshaft, refer to paragraph 3-30.1.

s. Remove flywheel and flywheel housing, refere to paragraph 3-31.1.

t. Remove screws (6), front plate (7), and gasket (8). Discard gasket.

CAUTION

Use care not to scratch crankshaft oil seal seating surfaces.

u. Remove rear oil seal and spacer.

v. Measure crankshaft end play at the center journal, refer to FIGURE 3-78. Do this before removing bearing caps. Standard end play is 0.004 in. (0.10mm), with a limit of 0.012 in. (0.30mm). If measured value exceeds limit, crankshaft thrust bearings (9) must be replaced.

NOTE

If crankshaft bearing cap bolts and bearings are to be reused, tag them as to their original locations to facilitate installation.

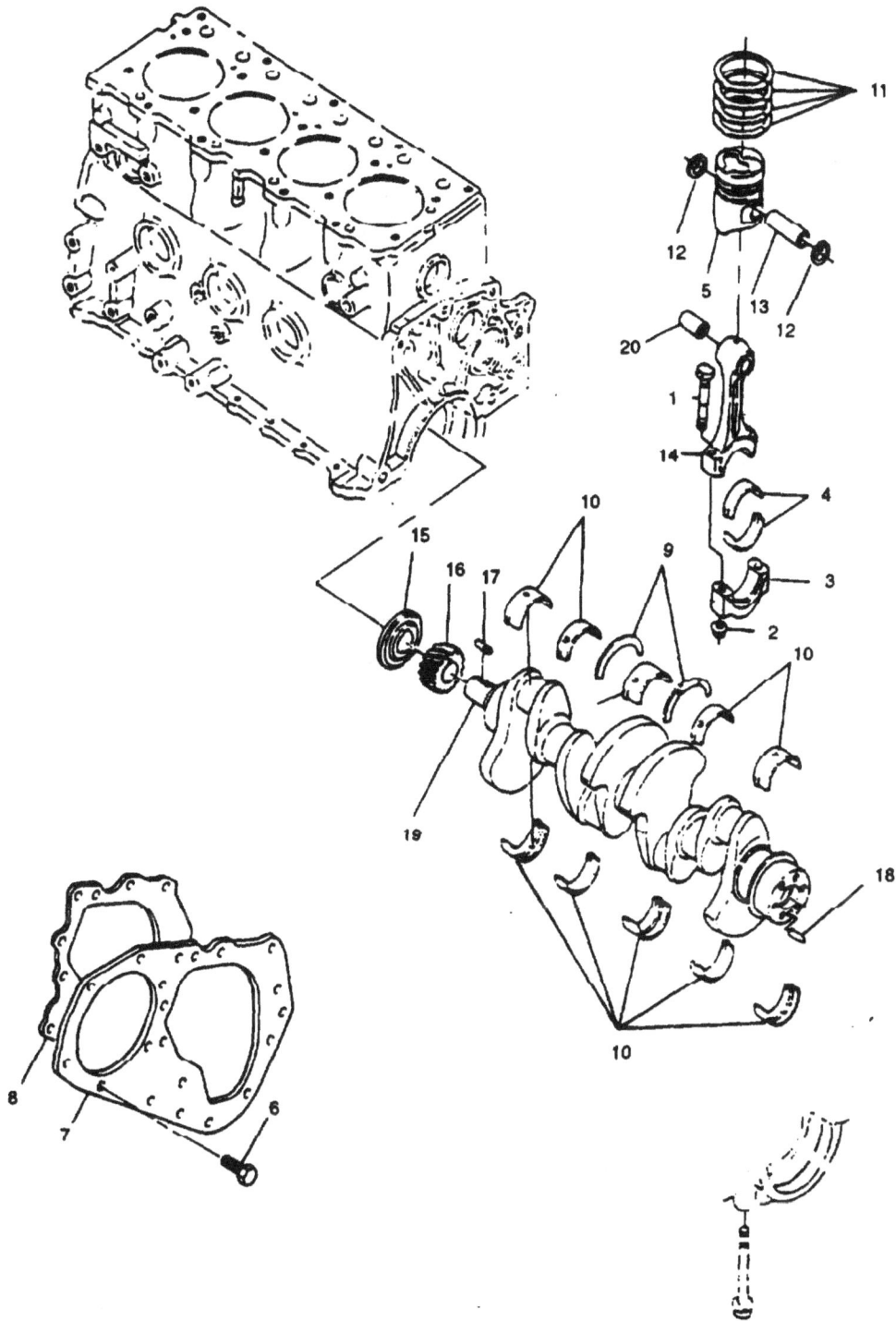

FIGURE 3-77. Pistons, Connecting Rods, and Crankshaft

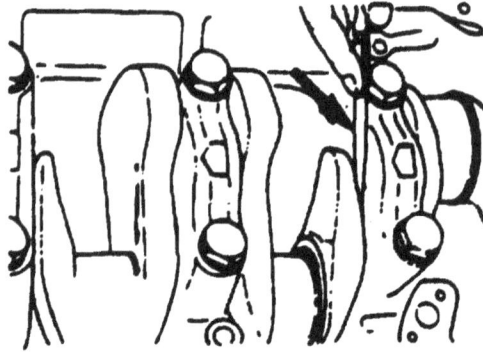

FIGURE 3-78. Crankshaft End Play

w. Loosen ten crankshaft baring cap bolts a little at a time in sequence shown in FIGURE 3-79.

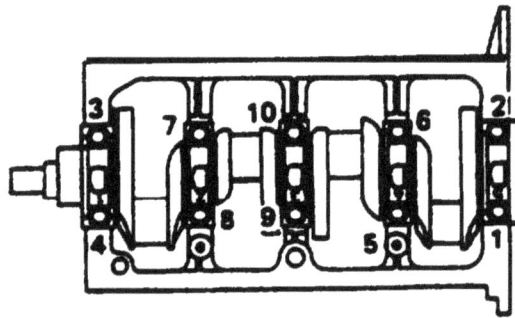

FIGURE 3-79. Crankshaft Bearing Cap Bolt Loosening Sequence

x.Remove ten bolts, five sets of bearings (10), two thrust bearings (9), and crankshaft from engine block.

3-28.2. Disassembly.

a Clamp piston connecting rod (14, FIGURE 3-77) in a softjawed vise.

CAUTION

Do not attempt to use some other tool to remove piston rings. Piston ring stretching will result in reduced piston ring tension.

b. Use a piston ring expander to remove four piston rings (11) from piston (5).

c. Remove two retainer rings (12) from piston pin (13).

NOTE

If piston and piston pin are to be reused, tag each piston and pin with cylinder number from which it was removed.

d. Using a hammer and a brass bar, tap piston pin (13) from piston (5)

e. Remove deflector (15) from crankshaft.

f. Using gear puller, remove crankshaft gear (16) from crankshaft. Remove key (17). 3-28.3.

Inspect and Measure

3-28.3.1.   Piston and Connecting Rod.

a. Check piston ring and piston ring groove clearance as follows:

(1) Using a feeler gage, measure the clearance between piston ring and piston ring groove at several points around the piston, refer to FIGURE 3-80.

(2) Standard clearance for the three compression rings is 0.0017 to 0.0027 in. (0.045 to 0.070 mm). Standard clearance for the oil ring is 0.0008 to 0.0021 in. (0.020 to 0.055 mm). The limit for clearance is 0.0059 in. (0.15 mm)

(3) If clearance exceeds limit, replace piston rings.

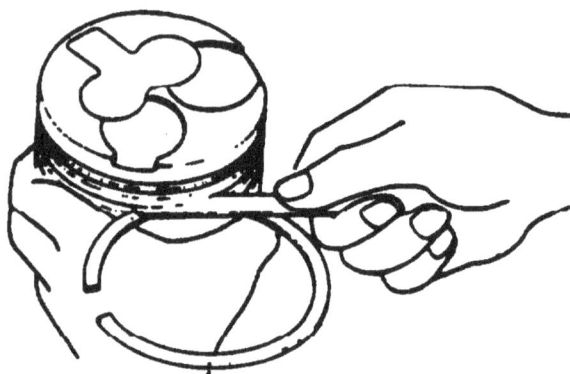

FIGURE 3-80. Piston Ring and Groove Clearance

b. Check piston ring gap as follows:

    (1) Insert a piston ring horizontally (in position as installed on piston), into cylinder liner bore, refer to FIGURE 3-81.

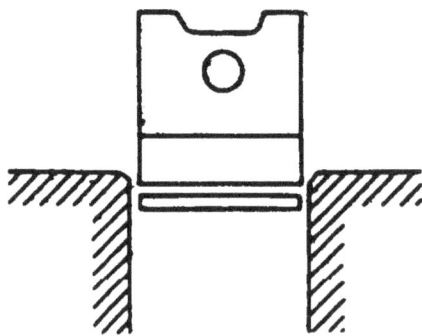

FIGURE 3-81. Piston Ring

(2) Using an inverted piston, push piston ring into cylinder liner approximately 0.39 in. (10 mm) or to the point where diameter of liner is smallest.

(3) Using a feeler gage, measure piston ring gap. Standard gap for three compression rings is 0.0079 to 0.0158 in. (0.2 to 0.4 mm). Standard gap for oil ring is 0.0039 to 0.0118 in. (0.1 to 0.3 mm) The limit for gap is 0.0591 in. (1.5 mm), refer to FIGURE 3-82.

(4) Inspect piston ring groove for damage or distortion. If damage or distortion is found, replace piston.

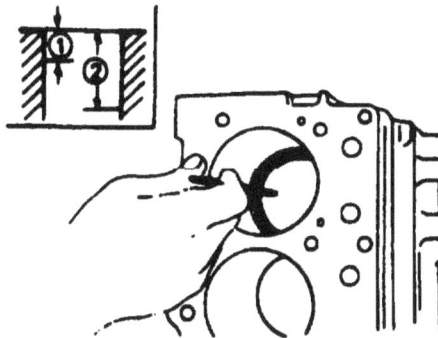

FIGURE 3-82. Measuring Piston Ring Gap

c. Using a micrometer, measure piston pin OD at several points. Standard OD is 1.063 in. (27.0 mm), with a limit d 1.062 in. (26.97 mm). If OD is less than limit, replace piston pm.

d. Check piston pin and piston clearance as follows:

(1) Use an inside micrometer to measure piston pin hole in piston.

(2) Stand& piston pin hole ID is 1.0525 to 1.0628 in. (26.938 to 25.997 mm). The clearance limit between pin and pin hole is 0.0 to 0.0002 in. (0.0 to 0.095 mm).

(3) If clearance exceeds limit, replace raplace piston.

(4) Use an inside micrometer to measure small end bushings (20, FIGURE 3-77). Replace bushings if ID is not within 1.0633 to 1.0635 in. (27.998 to 27.015 mm).

e. Check connecting rod side face clearance as follows:

(1) Install connecting rod on crankpin.

(2) Using a feeler gage, measure clearance between connecting rod side face and crankpin web face, refer to FIGURE 3-83.

(3) Standard clearance is 0.009 in. (0.23 mm), with a limit of 0.014 in. (0.35 mm).

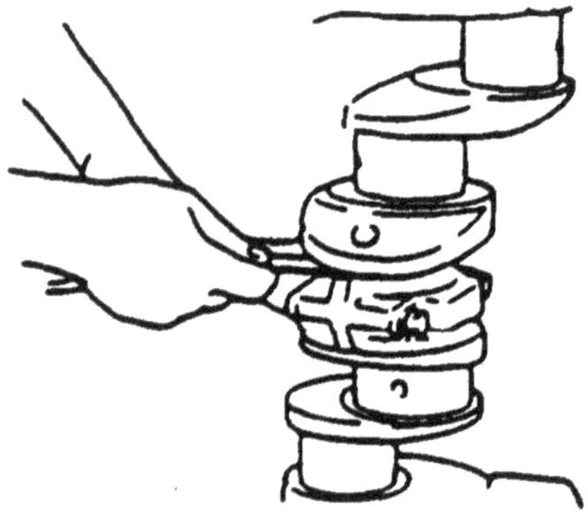

FIGURE 3-83. Connecting Rod Side Face Clearance

(4) If measured clearance exceeds limit, replace connecting rod.

3-28.3.2 <u>Crankshaft and Bearings.</u>

a. The following is a summation of basic dimension measurements, refer to FIGURE 3-84.

(1) Measure A and B dimensions, and subtract B from A to obtain dimension C.

(2) Measure crankpin to connecting rod bearing clearance D in same manner.

(3) Clearances C or D are basis for determining crankshaft repairs.

(4) Replace bearings with standard size ones when crankshaft journal wear and/or crankpin wear is within limits but bearing wear is close to wear limits.

(5) Replace bearings with undersized ones when crankshaft journal wear and crankpin wear exceeds limits.

b. Inspect crankshaft and bearings as follows:

(1)Inspect crankshaft journal surfaces and crankpin surfaces for excessive wear and damage.

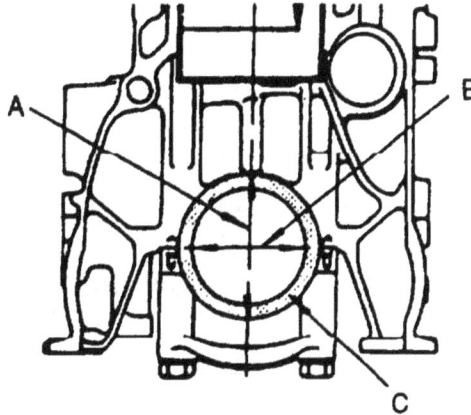

A - Main Bearing Inside Diameter

B - Crankshaft Journal Diameter

C - Crankshaft Journal to Main BearingClearance

FIGURE 3-84. Basic Dimension Measurements

(2)Inspect front and rear oil seal fitting surfaces of crankshaft for excessive wear and damage.

(3)Replace or repair crankshaft if any excessive wear or damage is found.

WARNING

Compressed air used for cleaning can create airborne particles that may enter the eyes. Pressure will not exceed 30 psig (207 kPa). Eye protection required.

(4)Inspect oil ports of crankshaft for obstructions. Use air pressure to clean oil ports.

c. Check crankshaft run-out as follows:

(1) Place crankshaft on V-blocks.

(2) Set an inside micrometer to center of crankshaft journal, refer to FIGURE 3-85.

(3) Gently rotatecrankshaft in normal direction of rotation.

(4) Read inside micrometer as you turn crankshaft. Standard run-out is 0.012 in. (0.03 mm) or less, with a limit of 0.0024 in. (0.06 mm)

(5) If measured run-out exceeds knit, replace crankshaft.

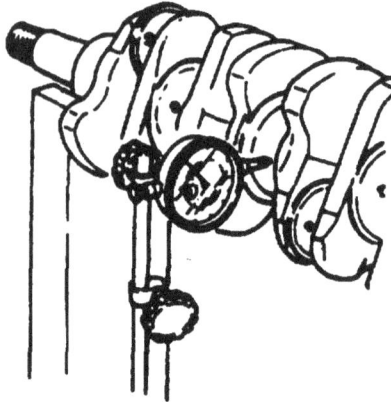

FIGURE 2-85. Crankshaft Run-out

d. Use a vernier caliper to measure spead, refer to FIGURE 3-86. The limit for bearing spread for crankshaft bearings is 2.93 in. (74.5 mm) and 2.22 in. (56.5 mm) for connecting rod bearing.

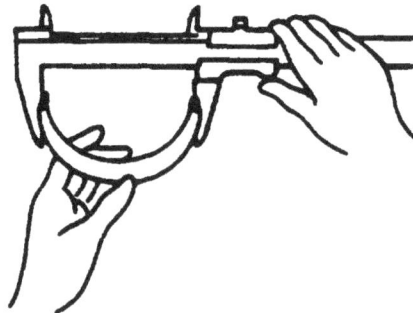

FIGURE 3-86. Bearing Spread Measurement

NOTE

If bearing gage (PLASTIGAGEPR1) method of checking clearance is used,
omit steps e, f, and g.

e. Check crankshaft journal and crankpin diameter as follows:

(1) Using a micrometer, measure crankshaft journal diameter at two places, 180 degrees apart and at two points along its length, refer to FIGURE 3-87.

(2) Repeat step (1) for crankpin.

(3) Standard diameter for crankshaft journal is 2.7548 to 2.7533 in. (69.920 to 69.632 mm), with a limit of 2.7560 in. (69.95 mm). Standard diameter for crankpin is 2.0849 to 2.0854 in. (52.918 to 52.930 mm), with a limit of 2.0862 in. (52.95 mm). Uneven wear standard value is 0.00004 in. (0.001 mm), with a limit of 0.002 in. (0.05 mm).

(4) If measured values are less than limits, crankshaft must be replaced or reground.

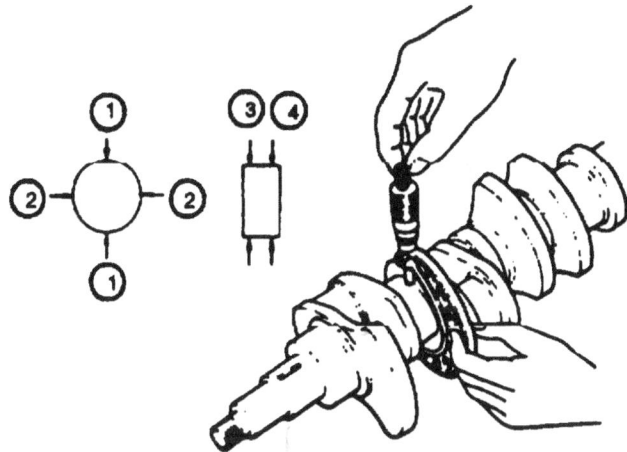

FIGURE 3-87. Crankshaft Journal and Crankpin Diameter

f. Check crankshaft journal and bearing clearance as follows:

(1) Install main bearings and bearing caps. Tighten bearing cap bolts to 123 ft-lbs (166.8 Nm).

(2) Using inside dial indicator, measure bearing inside diameter, refer to FIGURE 3-88.

(3) Standard clearance between bearing ID and crankshaft journal diameter (step e) is 0.0007 to 0.0025 in. (0.018 to 0.065 mm), with a limit of 0.0047 in. (0.12 mm).

(4) If clearance exceeds limit, replace bearing and/or crankshaft.

g. Check connecting rod bearing and crankpin clearance as follows:

(1) Install bearings on connecting rod.

FIGURE 3-88. Crankshaft Journal and Bearing Clearance

(2) Install bearing end cap and tighten bolts to 61.5 ft-bs (83.4 Nm).

(3) Using inside micrometer, measure bearing inside diameter, refer to FIGURE 3-89.

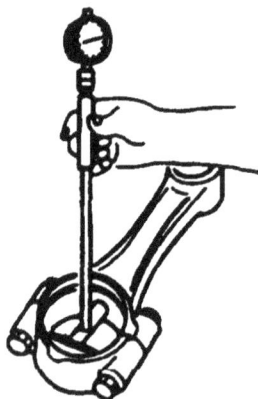

FIGURE 3-89. Connecting Rod Bearing Inside Diameter

(4) Standard clearance between bearing ID and crankpin (step e) is 0.0007 to 0.0025 in. (0.018 to 0.065 mm), with a limit of 0.0047 in. (0.12 mm)

(5) If clearance exceeds limit, replace bearings and/or crankshaft.

h. Check crankshaft journal and bearing clearance using bearing gage (PLASTIGAGEPRI) as follows:

(1) Clean cylinder block, journal bearing surfaces, bearing caps, and bearings.

(2) Install bearings in cylinder block.

(3) Carefully place crankshaft on bearings.

(4) Rotate crankshaft approximately 30 degrees to seat bearings.

(5) Place bearing gage over crankshaft journal across full width of bearing, refer to FIGURE 3-90.

NOTE

Do not allow crankshaft to turn during bearing cap installation and tightening.

(6) Install bearing caps with bearings. Tighten bearing cap bolts to 123 ft-lbs (166.8 Nm).

(7) Remove bearing caps.

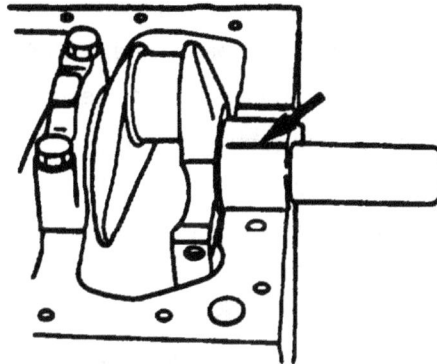

FIGURE 3-90. Crankshaft Journal and Bearing Clearance

(8) Compare width of bearing gage attached to either crankshaft or bearing against scale printed on bearing gage container, refer to FIGURE 3-91.

(9) Standard clearance is 0.0007 to 0.0025 in. (0.018 to 0.065 mm), with a limit of 0.0047 in. (0.12 mm).

(10) If measured clearance exceeds limit, perform clearance check in accordance with step f.

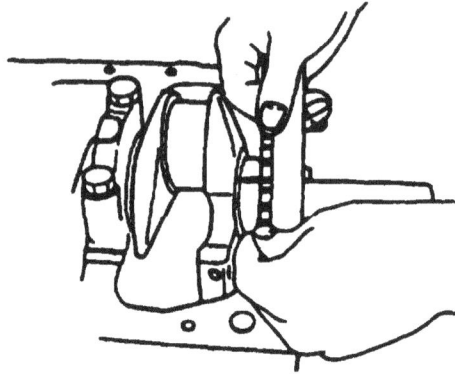

FIGURE 3-91. Comparing Bearing Gage

i. Check crankpin and bearing clearance using bearing gage (PLASTIGAGEPR1) as follows:

(1) Clean crankshaft, connecting rods, bearing cap, and bearings.

(2) Install bearing to connecting rod and bearing cap.

(3) Prevent connecting rod from moving.

(4) Apply engine lubricating oil (MIL-L-2104) to bearing gage to keep it from falling off and attach bearing gage to crankpin.

NOTE

Do not allow connecting rod and crankshaft to move when installing and tightening bearing cap.

(5) Install bearing cap and tighten bolts to 61.5 ft-bs (83.3 Nm).

(6) Remove bearing cap.

(7) Compare width of bearing gage attached to either crankshaft or bearing against scale printed on bearing gage container.

(8) Standard clearance is 0.0007 to 0.0025 in. (0.018 to 0.065 mm), with a limit of 0.0047 in. (0.12 mm).

(9) If measured clearance exceeds limit, perform clearance check in accordance with step g.

3-28.4. Assembly.

NOTE

Crankshaft gear timing mark 'X-X must face out.

a. Install key (17, FIGURE 3-77) on crankshaft and using gear installer (320908) install gear (16) on crankshaft, refer to FIGURE 3-92.

b. Install deflector (15, FIGURE 3-77) on crankshaft.

c. Clamp connecting rod in a soft-jawed vice.

FIGURE 3-92. Crankshaft Gear Installation

d. Install one retaining ring (12) in one side of piston (5).

NOTE

If a heater is not available, use a hot plate to heat a container of oil to 176°F (80°C). Place piston in hot oil.

e. Place piston (5) in a heater and heat piston to 176°F (80°C).

WARNING

Use protective gloves when removing pistons from heater to avoid seri-ous burns.

f. Apply coat of engine lubricating oil (ML-L-21 04) to piston pin (13).

g. Position heated piston on connecting rod ensuring piston head front mark (notch on top of piston) and connecting rod ISUZU forging mark are facing same direction.

h. Using your hands and fingers, push piston pin (13) into piston until it makes contact with retaining ring.

i. Install other retaining ring (12) in piston.

j. Using a piston ring expander, install piston rings (11) as shown in FIGURE 3-93 and the following:

NOTE

Install compression rings with stamped side facing up.

NOTE

Insert expander coil into oil ring groove so there is no gap on either side of expander coil before installing oil ring.

(1) Install piston rings (11, FIGURE 3-77) in order shown in FIGURE 3-93.

(2) Apply engine lubricating oil (ML-L-21 04) to piston ring surface.

(3) Check that piston rings slide smoothly in piston ring grooves.

(4) Oil ring coil expander connecting portion must not have a gap.

k. Repeat steps c thru j for remaining pistons and connecting rods.

FOUR RINGS TYPE

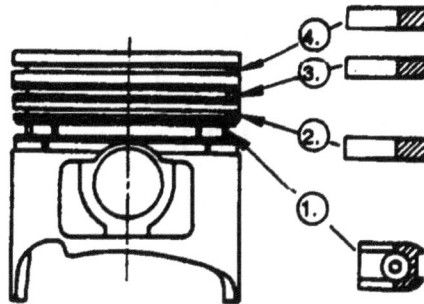

1.   Oil Ring

2.   First Compression Ring

3. Second Compression Ring

4. Third Compression Ring**FIGURE 3-93. Piston Ring Installation**

3-28.5. Installation.

a. Carefully wipe any foreign material from upper bearings.

b. If crankshaft bearings are being reused, locate position marks applied during removal.

c. Refer to FIGURE 3-94 for correct bearing location at cylinder block bearing fitting arches and bearing caps.

d. Fit bearing tang firmly into slot machined on cylinder block bearing arches.

NOTE

Do not apply oil to bearing back faces and cylinder block bearing fiitting surfaces.

e. Apply an ample coat of engine lubricating oil (MIL-L-2104) to crankshaft journals and crankshaft hearing

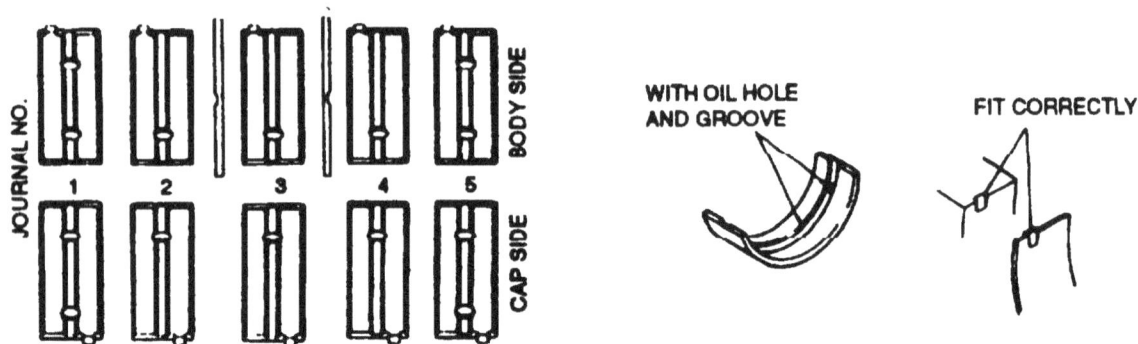

sur-faces.

FIGURE 3-94. Crankshaft Bearing Location

f. g. Install hearing halves (10, FIGURE 3-77) on engine block and install crankshaft (19).

h. Apply an ample coat of engine lubricating oil (MIL-L-2104) to thrust hearings (9) and install them to crankshaft center journal with grooves facing crankshaft web.

i. Install lower halves of hearings (10) in main bearing cap and rod cap (3).

Perform following for bearing cape No. 1 and No. 5.

(1) Apply sealing compound (FORMAGASKET2) to inside surfaces of new arch gaskets.

(2) Use your fingers and push arch gasket into groove of No. 1 and No. 5 hearing caps.

NOTE

Use care not to scratch outer surface of arch gasket.

j. Apply a coat of silicon adhesive (564R46001101) to cylinder block fitting surface for No. 1 and No. 5 bearing cap.

k. Install No. 1 and No. 5 bearing caps and ensure arch gasket makes firm contact with cylinder block side.

l. Install remaining bearing caps with bearing cap head arrow mark facing forward. Bearing cap stamped with a letter (A, B, C, etc.) must be installed at the No. 2 position.

m. Apply a coat of engine lubricating oil (MIL-L-2104) to ten bearing cap bolts and install them in bearing caps hand-tight.

n. Tighten bearing cap bolts to 123.0 ft-lbs (166.6 Nm), a little at a time, in sequence shown in FIGURE 3-95.

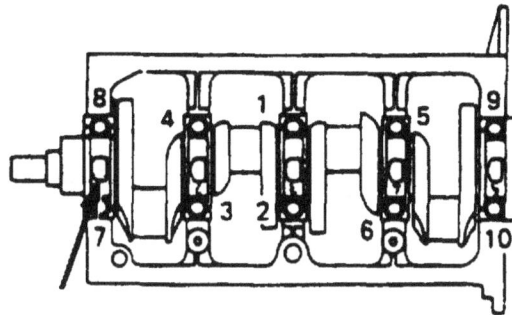

FIGURE 8-95. Crankshaft Bearing Cap Bolts Tightening Sequence

o. Ensure crankshaft turns smoothly by rotating it manually.

p. Install rear oil seal (1, FIGURE 3-96) and spacer (2) as follows:

(1) Apply engine lubricating oil (MIL-L-2104) to oil seal (1) lip circumference.

(2) Apply silicon adhesive (564R46001101) to oil seal outer circumference.

(3) Place spacer (2) on oil seal (1).

(4) Using crankshaft rear oil seal installer (NU7627), install oil seal (1) cylinder block.

(5) Check that no gap exists between spacer (2) and oil seal (1).

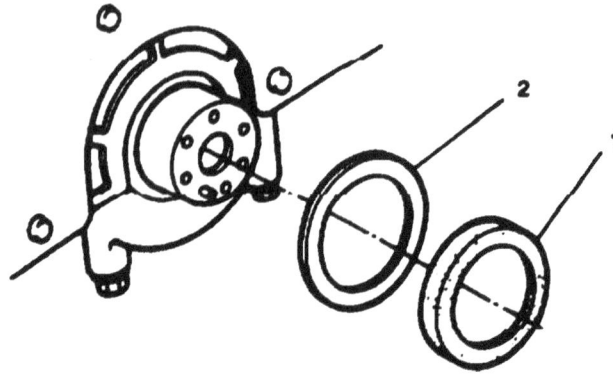

FIGURE 3-96. Crankshaft Rear Oil Seal and Spacer

q. Install front plate (7, FIGURE 3-77) with new gasket (8) and secure with screw (6). Tighten screws to 168 in-lbs (19.0 Nm).

r. Install flywheel and flywheel housing, refer to paragraph 3-31.3.

s. install camshaft, refer to paragraph 3-30.5.

t. Apply a coat of engine lubricating oil (MIL-L-2104) to circumference of each piston ring (11) and piston (5).

NOTE

Position the piston ring 1800 from each other at the top. The piston ring gap must be placed at the piston pin side having no thrust pressure.

u. Position piston ring gaps as shown in FIGURE 3-97.

v. Apply a coat of motybdenum disulfide grease (MOLYKOTE Z) to the two piston skirts. This will facilitate smooth break-in when engine is first started after reassembly.

w. Apply a coat of engine lubricating oil (MIL-L-2104) to upper bearing surfaces and cylinder bore.

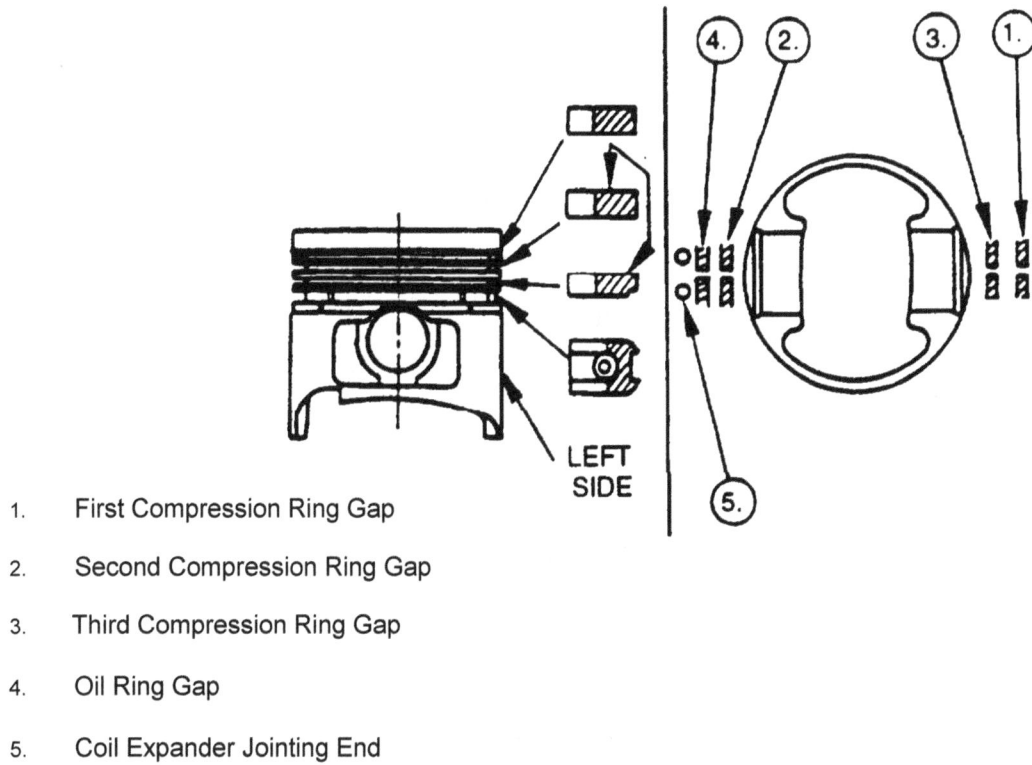

1. First Compression Ring Gap

2. Second Compression Ring Gap

3. Third Compression Ring Gap

4. Oil Ring Gap

5. Coil Expander Jointing End

FIGURE 3-97. Piston Ring Gap Setting

CAUTION

Ensure rod bolts do not come in contact with crankpin. Damage to crankpin could result.

x. Using a piston ring compressor, compress piston rings. Use a hammer handle to push piston into cylinder bore until connecting rod makes contact with crankpin. At same time rotate crankshaft until crankpin is at BDC.

y. Position piston head front mark so it faces front of engine.

z. Align bearing cap (3, FIGURE 3-77) cylinder number marks and connecting rod cylinder number marks. The cylinder number marks must be turned toward exhaust manifold.

aa.Apply a coat of engine lubricating oil (MIL-L-2104) to threads and seating faces of connecting rod cap bolts (1). Install two bolts (1) and nuts (2) for each beating cap (3). Tighten nuts to 61.5 ft-lbs (83.3 Nm).

ab. Ensure crankshaft rotates freely.

ac. Install oil pump, refer to paragraph 3-18.3. ad. Install crank-

case, refer to paragraph 3-29.3. ae. Install cylinder head, refer to

paragraph 3-27.6. af. Install fuel injection pump, refer to para-

graph 3-11.6. ag. Install idler gear and shaft, refer to paragraph

3-30.5. ah. Install front cover, refer to paragraph 3-25.3. ai.Install

oil pan, refer to paragraph 3-17.3. aj. Install PCV system, refer

to paragraph 3-20.3. ak. Install water pump, refer to paragraph

3-7.4. al. Install alternator, refer to paragraph 3-24.6. am. Install

starter, refer to paragraph 3-23.6.

an. Install fuel filter assembly, refer to paragraph 3-9.4. ao.

Install oil filter, refer to paragraph 3-14.3. 3-29.___CRANKCASE.

3-29.1. Removal.

   a. Remove oil pan, refer to paragraph 3-17.1.

   b. Remove two screws (1, FIGURE 3-98) securing crankcase to flywheel housing.

   c. Remove two bolts (2) two bolts (3) two screws (4) and fourteen screws (5) securing crankcase (6) to engine block.

   d. Remove crankcase (6) from engine block using a screwdriver or pry bar in slit provided on crankcase mounting flange. Remove two shims (7).

3-29.2. Inspection.

   a. Inspect crankcase for cracks, deformities, or other damage. If found, replace crankcase.

   b. Remove any traces of sealing compound (FORMAGASKET2) from mating surfaces of crankcase and engine block.

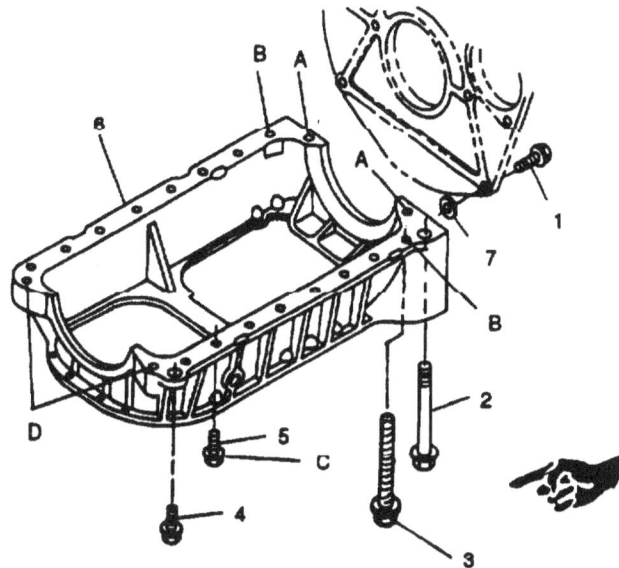

A - 100 mm bolt
B - 120 mm bolt / lockwasher
C - 22 mm screw
D - 45 mm bolt

FIGURE 3-98. Crankcase

3-29.3. Installation.

a. Apply a coat of sealing compound (FORMGASKET2) to engine block mating surface and to No. 1 and No. 5 bearing cap arch gaskets.

b. Position crankcase (6, FIGURE 3-98) on engine block, aligning dowel pin in block with hole in crankcase. Secure crankcase to engine block with fourteen screws (5) two bolts (2) two bolts (3) and two screws (4).

c. Install two shims (7) and two screws (1) securing crankcase to flywheel housing.

d. Tighten all fasteners a little at a time and in sequence shown in FIGURE 3-99 to 168 in-lbs (19.0 Nm).

e. Install oil pan, refer to paragraph 3-17.3.

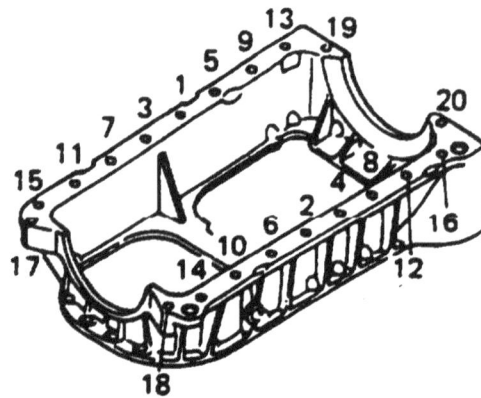

FIGURE 3-99. Crankcase Fastener Tightening Sequence

3-30. <u>CAMSHAFT AND IDLELR GEAR ASSEMBLIES.</u> 3-30.1.

<u>Removal.</u>

    a. Remove oil filter, refer to paragraph 3-14.2.

    b. Remove fuel fitter, refer to paragraph 3-9.1.

    c. Remove starter, refer to paragraph 3-23.1.

    d. Remove attemator, refer to paragraph 3-24.1.

    e. Remove water pump, refer to paragraph 3-7.1.

    f. Remove PCV system, refer to paragraph 3-20.1.

    g. Remove oil pan, refer to paragraph 3-17.1.

    h. Remove front gear cover, refer to paragraph 325.1.

    i.   Measure timing gear backlash (refer to FIGURE 3-100) as follows:

      (1) Set a dial indicator on timing gear to be measured.

(2) Hold gear to be checked and adjoining gear stationary.

(3) Move gear tooth side to side in backlash. Record indicated backlash TIR. Standard TIR is 0.004-0.007 in. (0.10 to 0.17 mm), with a limit of 0.012 in. (0.30 mm).

(4) If TIR exceeds limit, replace gear.

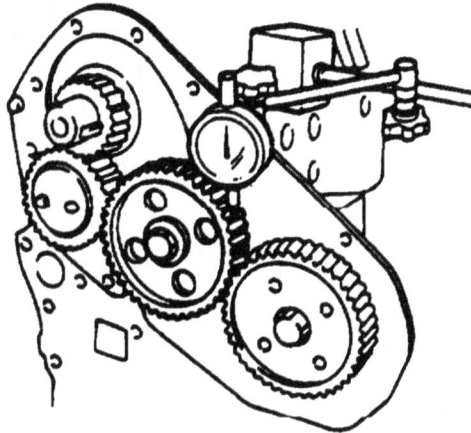

FIGURE 3-100. Measuring Timing Gear Backlash

j. Measure idler gear end play as follows:

(1) Insert feeler gage between idler gear (3, FIGURE 3-101) and thrust collar (2) and measure gap.

(2) Standard end play (gap) is 0.003 in. (0.07 mm), with a limit of 0.008 in. (0.2 mm)

(3) If end play exceeds limit, replace thrust collar (2).

k. Remove two screws (1) securing thrust collar (2), idler gear (3) and shaft (4). Remove gear (3) from shaft (4).

l. Remove fuel injection pump, refer to paragraph 3-11.1.

J. Remove cylinder head, refer to paragraph 3-27.1.

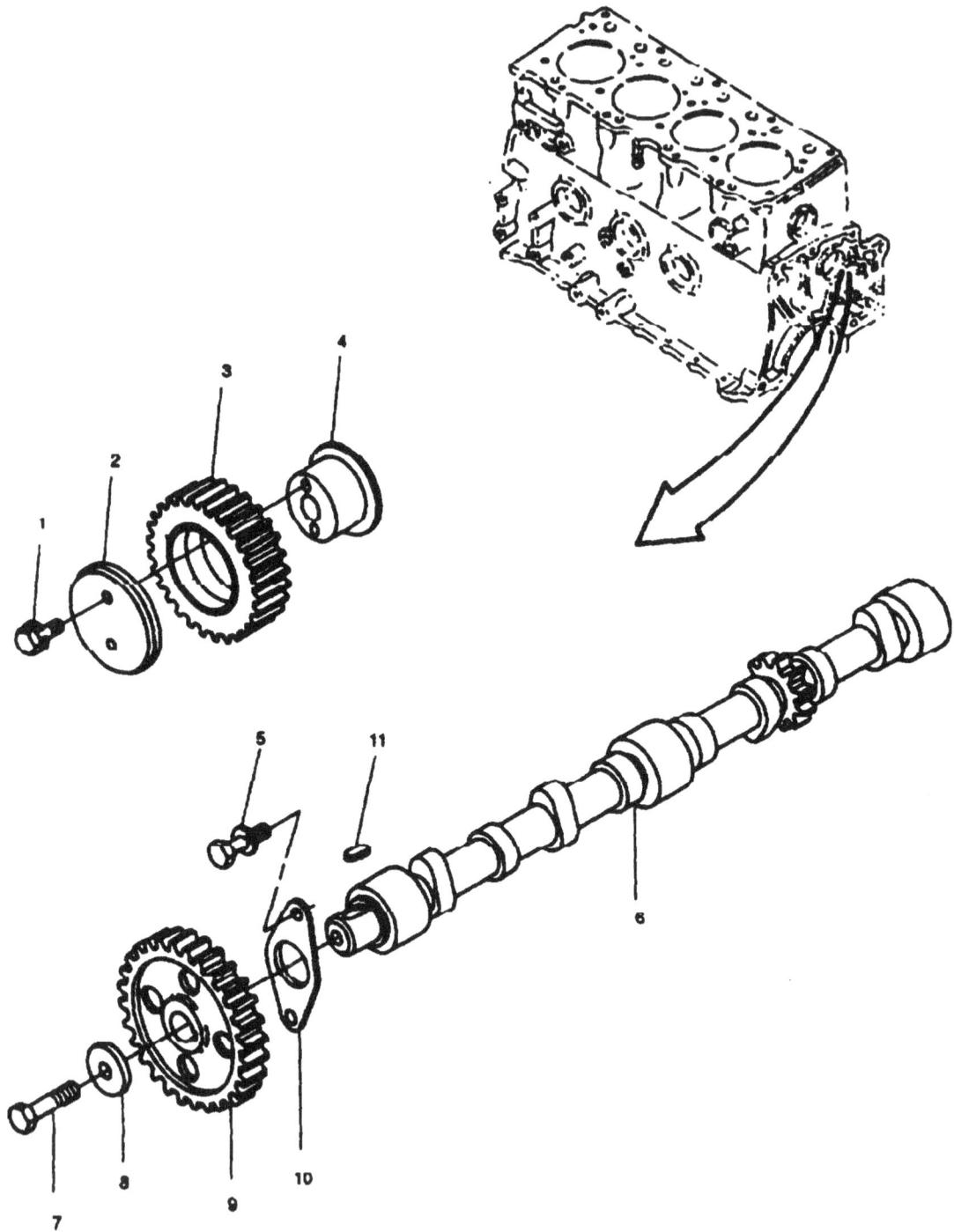

FIGURE 3-101. Camshaft and Idler Gear Assemblies

n. Remove crankcase, refer to paragraph 3-29.1.

o. Remove oil pump, refer to paragraph 3-18.1.

p. Remove pistons and connecting rods, refer to paragraph 3-28.1

q. Remove two screws (5) at camshaft thrust plate (10).

> ┌─────────────┐
> │ **CAUTION** │
> └─────────────┘
>
> When removing camshaft (6) use extreme care to avoid contact be-tween camshaft and engine block. Remove it slowly and use both hands, otherwise, equipment damage could occur.

r. Withdraw camshaft (6) from engine block using care not to damage camshaft journal, cam, and camshaft.

3-30.2. Disassembly of Camshaft.

a. Clamp camshaft (6, FIGURE 3-101) in a soft-jawed vise.

b. Measure camshaft end play, refer to paragraph 3-30.3.2, step f.

c. Remove bolt (7) and washer (8) from gear end of camshaft (6)

d. Using a universal puller, remove gear (9) from camshaft (6)

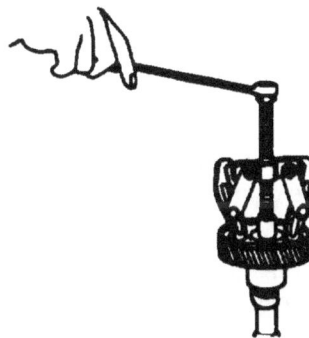

FIGURE 3-102. Thrust Plate Removal

3-30.3. Inspect and Measure. 3-30.3.1.

Idler Gear and Shaft.

a. Using a micrometer, measureidler gear shaft OD, refer to FIGURE 3-103. standard 00 is 1.769 to 1.770 in. (44.95 to 44.98 mm), with a limit of 1.767 in. (44.96) If OD is less than limit, replace shaft.

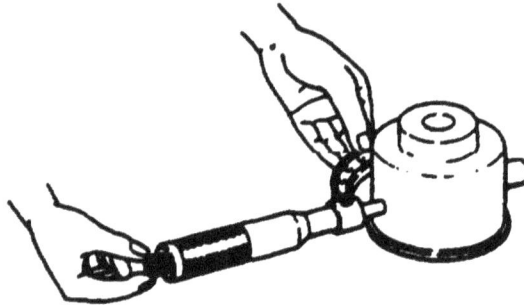

FIGURE 3-103. Idler Gear Shaft Outside Diameter

b. Using inside micrometer, measure idler gear ID. Standard ID is 1.7717 to 1.7718 in. (45.0 to 45.03 mm), with a limit of 1.7756 in. (45.10 mm).

c. Standard clearance between idler gear shaft OD and idler gear ID is 0.0010 to 0.0033 in. (0.025 to 0.085 mm) . If clearance exceeds limit of 0.608 in. (02 mm), replace idler gear.

3-30.3.2. camshaft.

a. Inspect journals, cams, oil pump drive gear, and camshaft bearing for excessive wear and damage. If excessive wear or damage is found, replace camshaft and bearings.

b. Standard camshaft journal diameter is 1.88 to 1.89 in. (47.94 to 47.97 mm). Using a micrometer, measure journal diameter two places, 180 degrees apart, refer to FIGURE 3-104. If measured diameter is less than limit of 1.87 in. (47.6 mm), replace camshaft.

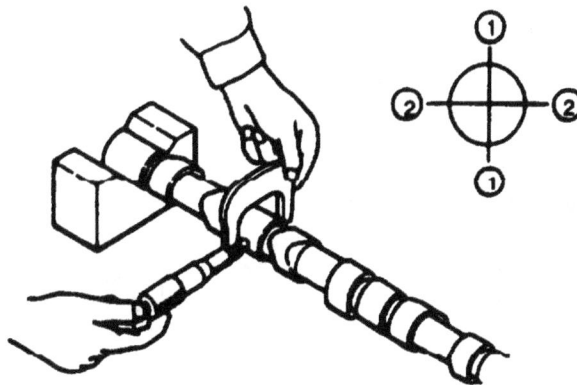

FIGURE 3-104. Camshaft Journal Diameter

c. Standard cam height is 1.60 in. (40.57 mm). Using a micrometer, measure cam height, refer to FIGURE 3-105. If cam height is less than limit of 1.58 in. (40.2 mm), replace camshaft.

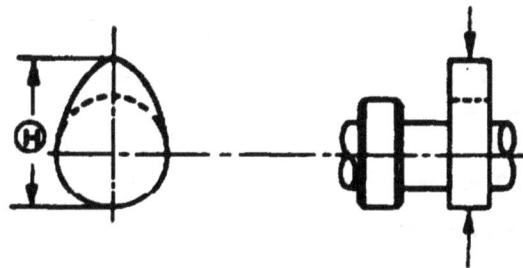

FIGURE 3-105. Cam Height

d. Check camshaft run-out as follows:

(1) Placecamshaft on V-blocks.

(2) Measure run-out with a dial indicator, refer to FIGURE 3-106.

(3)Standard run-out is 0.002 in. (0.05 mm). If runout exceeds limit of 0.004 in. (0.10 mm), replace camshaft.

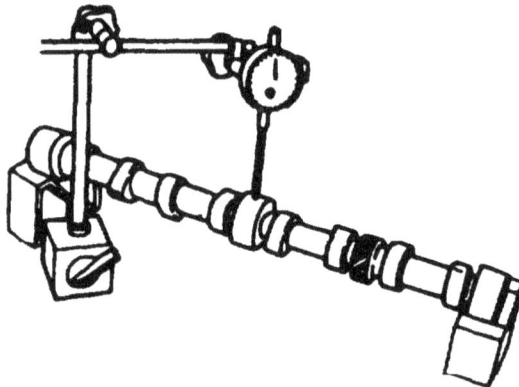

FIGURE 3-106. Camshaft Runout

e. Check camshaft and camshaft bearing clearance as follows:

(1) Use inside dial indicator and measure camshaft bearing ID.

(2) Standard bearing ID is 1.8897 to 1.8909 in. (48.0 to 48.03 mm), with a limit of 1.8929 in, (48.08 mm).

(3) Standard clearance between camshaft journal OD (measured in step b), and camshaft bearing ID is 0.002 in. (0.05 mm). If clearance exceeds limit of 0.005 in. (0.12 mm), replace bearings. Refer to paragraph 3-311.1.

f. Before disassembling camshaft, check camshaft end play as follows:

(1) Push thrust plate as far as it will go toward camshaft gear.

(2) Use a feeler gage and measure clearance between thrust plate and camshaft journal.

(3) Standard clearance (end play) is 0.082 to 0.0044 in. (0.080 to 0.114 mm). If clearance exceeds limit of 0.008 in. (0.2 mm), replace thrust plate.

3-30.4. Assembly of Camshaft.

    a. Clamp camshaft (6, FIGURE 3-101) in a soft-jawed vise.

    b. install key (11) and thrust plate (10) on camshaft (6).

    c. Place timing gear (9) on camshaft with stamped timing mark facing out.

    d. Apply engine lubricating oil (ML-L-2104) to threads of bolt (7). Install bolt (7) and washer (8) on camshaft and tighten bolt to 79.5 ft-lbs (107.8 Nm).

3-30.5. Installation.

    a. Apply a coat of engine lubricating oil (ML-L-2104) to camshaft (6, FIGURE 3-101) and camshaft bearings.

> **CAUTION**
>
> When installing camshaft (6) use extreme care to avoid contact be-tween camshaft and engine block. Install it slowly and use both hands. Otherwise, equipment damage could occur.

    b. Slide camshaft (6) into engine block, using care not to damage camshaft bearings.

    c. Secure thrust plate (10) to engine block using two screws (5). Tighten screws to 168 in-lbs (19.0 Nm)

    d. Manually rotate camshaft and check for smooth rotation.

    e. Install pistons and connecting rods, refer to paragraph 3-28.5.

    f. Install oil pump, refer to paragraph 3-18.3.

    g. Install crankcase, refer to paragraph 3-29.3.

    h. i. Apply a coat of engine lubricating oil (ML-L-2104) to idler gear (3) and idler gear shaft (4) bearing.

    Position idler gear shaft (4) and idler gear (3) on engine block with setting marks X and "Y facing front of engine.

j. k.

Align idler gear setting mark X with crankshaft timing gear setting mark X-X".

Align idler gear setting mark "Y" with camshaft timing gear setting mark "Y-Y".

l.

Install thrust collar (2) on idler gear and shaft and secure with two screws (1). Tighten screws to 168 in-lbs (19.0 Nm).

m. Install cylinder head, refer to paragraph 3-27.6.

n. o. Install fuel injector pump, refer to paragraph 3-11.6.

p. Install front gear cover, refer to paragraph 3-25.3.

q. Install oil pan, refer to paragraph 3-17.3.

Install PCV system, refer to paragraph 3-20.3.

r. Install water pump, refer to paragraph 3-7.4.

s. Install thermostat housing, refer to paragraph 3-6.4.

t. Install alternator, refer to paragraph 3-24.5.

u. Install flywheel housing, refer to paragraph 3-31.3.

v. Install starter, refer to paragraph 3-23.6.

w. Install fuel filter, refer to paragraph 3-9.4.

x. Install oil filter, refer to paragraph 3-14.3. 3-31.

FLYWHELL AND HOUSING. 3-31.1. Removal.

a. Remove starter, refer to paragraph 3-23.1.

b. Block flywheel (3, FIGURE 3-107) with a piece of hard wood.

c. Remove four bolts (1) and washers (2) securing flywheel assembly (3) to basic flywheel (6). Remove flywheel assembly (3)

d. Loosen six flywheel attaching bolts (4) a little at a time and in sequence shown in FIGURE 3-108.

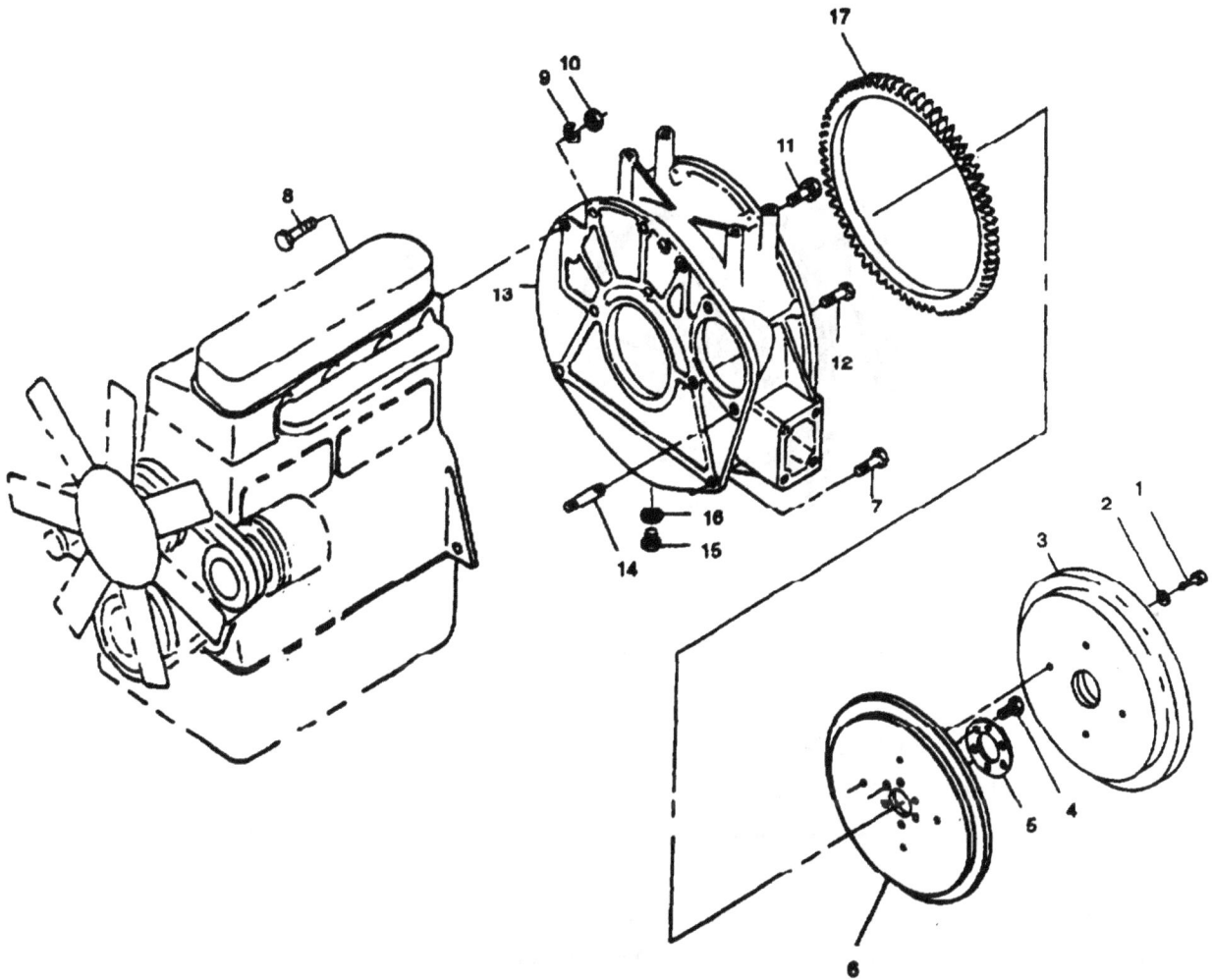

FIGURE 3-107. Flywheel and Housing

e. Remove six bolts (4, FIGURE 3-107), flywheel washer (5), and basic flywheel (6) from crankshaft.

f. Remove two bolts (7) securing flywheel housing to crankcase.

g. Remove (8), lockwasher (9), nut (10), three bolts (11), and three bolts (12) securing flywheel housing (13) to engine block. Remove flywheel housing. Discard lockwasher (9).

h. If necessary, remove two starter mounting studs (14), plug (15) and gasket (16). Discard gasket (16).

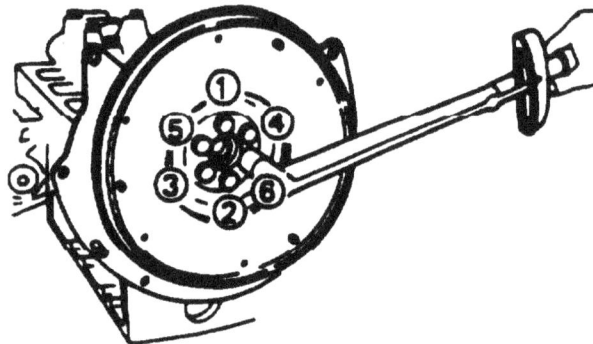

FIGURE 3-108. Flywheel Bolt Loosening Sequence

3-31.2. Inspection/Repair.

a. Inspect ring gear (17, FIGURE 3-107) for broken, missing, or excessively worn gear teeth.

b. Inspect flywheel housing (13) for cracks or other damage.

c. Remove all traces of sealing compound (FORMAGASKET2) from mating surface of flywheel housing and engine block.

d. If ring gear (17) is damaged, replace it as follows:

(1) Using a hammer and chisel, strike around edges of ring gear (17) to remove it from basic flywheel (6).

```
WARNING
```

Use protective gloves when handling heated ring gear to prevent seri-ous burns.

(2) Evenly heat ring gear to not more than 390°F (200°C). This will cause ring gear to expand.

(3) Install heated ring gear (17) with chamfer facing rear of engine on basic flywheel (6).

3-31.3. Installation.

    a. If removed, install plug (15, FIGURE 3-107) with new gasket (16) in flywheel housing and install two starter mounting studs (14).

    b. Apply sealing compound (FORMAGASKET2) to shaded areas at flywheel housing (10) shown in FIGURE 3-109.

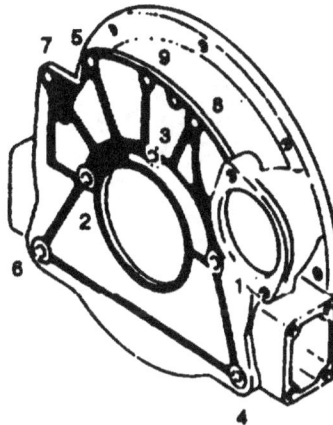

FIGURE 3-109. Flywheel Housing Installation Sequence

    c. Position flywheel housing on engine block and secure with three bolts (12, FIGURE 3-107), three bolts (11), bolt (8). new lockwasher (9), and nut (10). Install two bolts (7) securing flywheel housing to crankcase.

    d. Tighten flywheel housing fasteners a little at a time and in sequence shown in FIGURE 3-109 to 61 ft-lbs (82.7 Nm) for bolts (12) and to 30 ft-lbs (40.7 Nm) for bolts (11).

e. Position basic flywheel (6) on crankshaft, aligning dowel pin.

f. Coat threads of six flywheel attaching bolts (4) with engine lubricating oil (ML-L-2104).

g. Install flywheel washer (5) on crankshaft and install six bolts (4). Tighten bolts to 79.5 ft-lbs (107.9 Nm) in sequence shown in FIGURE 3-110.

NOTE

Use a block of hard wood to prevent flywheel from turning while tighten bolts.

h. Position flywheel assembly (3) on basic flywheel (6) and secure with four bolts (1) and washers (2).

i. Install starter, refer to paragraph 3-23.6.

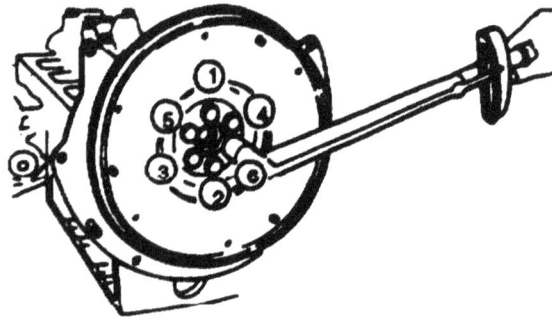

FIGURE 3-110. Flywheel Bolt Tightening Sequence

3-32. CYCLINDER BLOCK ASSEMBLY.

3-32.1. Disassembly.

a. Remove crankshaft, refer to paragraph 3-28.1.

b. Remove bolts (1, FIGURE 3-111) and washers (2) securing left engine foot (3) to cylinder block; remove foot.

c. Remove bolts (4) and washers (5) securing right engine foot (7) and alternator mounting bracket (6); remove foot and bracket.

d. Remove small freeze plug (8) on rear of engine and using a punch, drive coolant duct (9) from cylinder block.

e. Remove eight tappets (10) from block. If they are to be reused, tag them as to their location for use during assembly.

NOTE

To remove rear camshaft bearing, remove freeze plug.

f. Remove camshaft bearings (11) using bearing remover/ installer (320905)

g. Remove two screws (12) securing cover (13) to engine block; remove cover (13) and packing (14). Discard packing.

NOTE

If cylinder liner is removed, it must be replaced.

h. Remove cylinder liner (15) as follows:

(1) Insert cylinder liner remover (NU-7626) into cylinder block (from lower side) until it makes firm contact with cylinder liner. Insert grip (7631) into cylinder liner remover.

```
CAUTION
```

Use care not to damage cylinder block upper face during cylinder liner removal.

(2) Using an arbor press, slowly force cylinder liner from cylinder block.

i. Remove plugs, caps, and freeze plugs only if necessary.

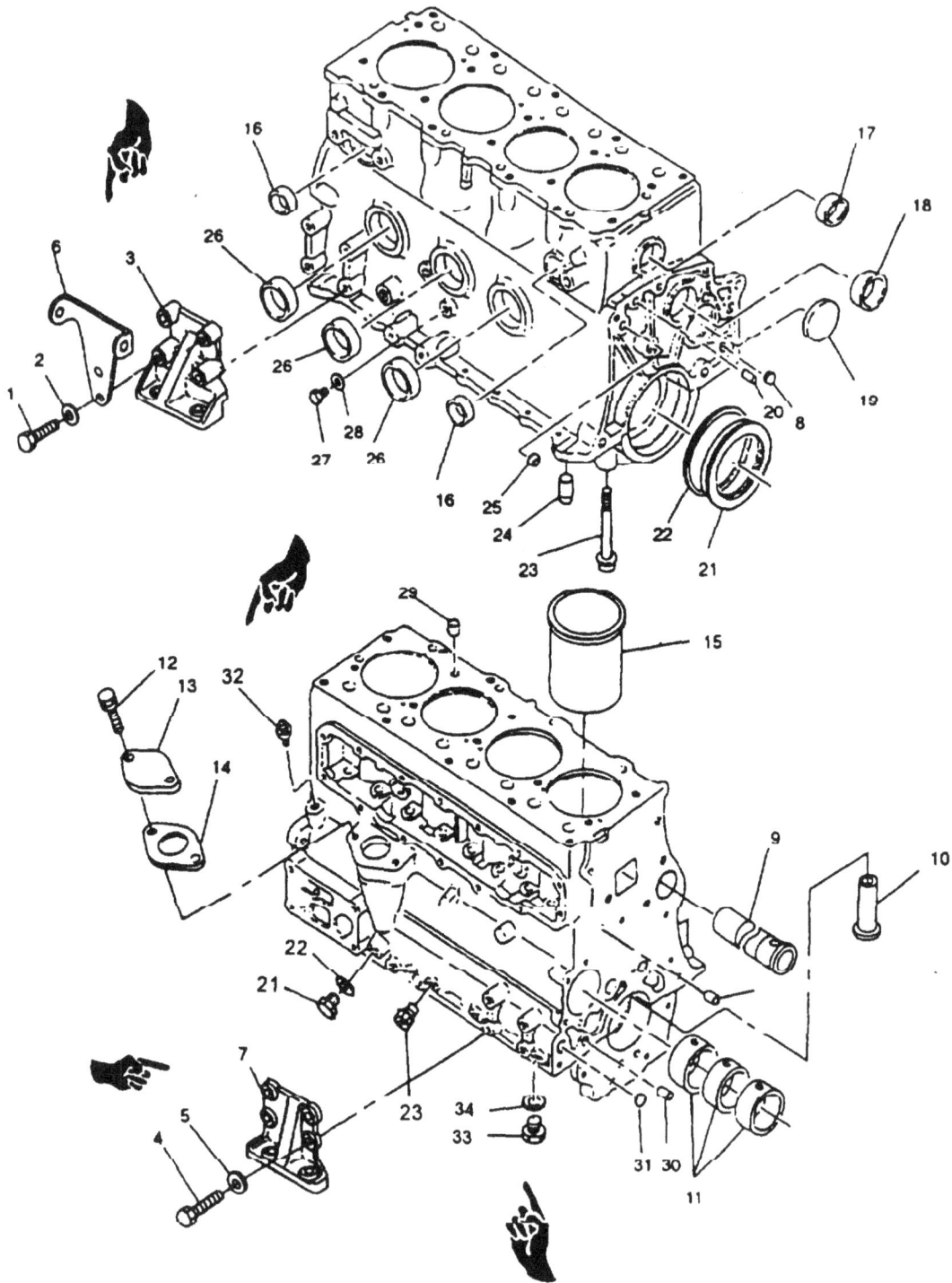

FIGURE 3-111. Cylinder Block Assembly

j.   Remove dowel pins and fittings only if necessary.

3-32.2. <u>Inspection/Repair.</u>

NOTE

Letters and numbers are stamped on top face of block at each cylinder bore.Letters (A or C) represent grade of piston installed during
production. Numbers (1, 2, 3, or 4) represent grade of cylinder liner
installed during production. However. when ordering a new cylinder liner
for service repair either a No. 2 or No. 3 grade cylinder liner will be
supplied. Either one may be used to replace whatever grade liner was
used in production.

(1)   If not removed, remove dowel pin.

(2)   Remove cylinder liner, refer to paragraph 3-32.1 step h.

(3)   Using a straight edge and a feeler gage, measure the four sides and two diagonals of cylinder
block upper face, refer to FIGURE 3-112.

(4)   Standard warpage is 0.002 in. (0.05 mm) or less. If measured value exceeds 0.008 in. (0.2 mm)
but is less than 0.016 in. (0.4 mm), regrind the cylinder block. If measured value is greater
than 0.006 in. (0.14 mm), replace cylinder block.

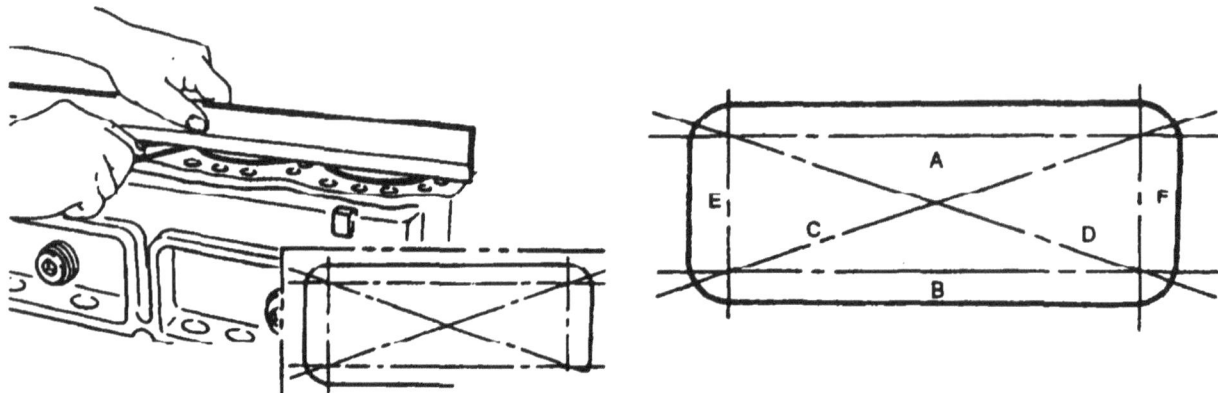

FIGURE 3-112. Cylinder Block Upper Face Warpage Measurement

(5) Standard cylinder block height is 10.748 in. (273 mm) with a limit of 10.732 in. (272.6 mm), refer to FIG-URE 3-113.

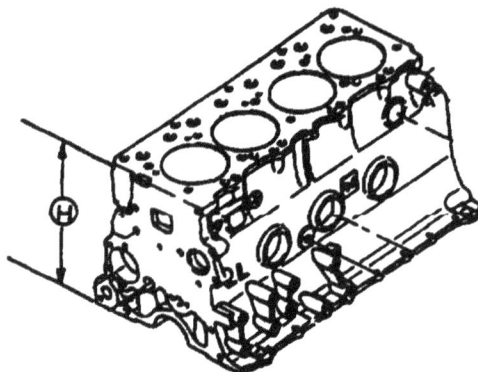

FIGURE 3-113. Cylinder Block Height Measurement

(6) Install cylinder liner, refer to paragraph 3-32.3.

(7) Install cylinder bled< dowel pin. b.Mea-

sure cylinder liner bore as follows:

(1) Using a cylinder indicator, measure cylinder liner bore at a point 0.59 to 0.75 in. (15 to 19 mm) from top of liner where most wear is present. Take measurement in both the thrust and axial directions of the crank-shaft.

(2) Standard cylinder liner bore is 3.386 in. (86.0 mm) with a limit of 3.394 in. (86.2 mm) If bore exceeds limit, replace cylinder liner.

NOTE

Inside of dry type cylinder liner is chrome plated and cannot be rebored.

NOTE

If inside of cylinder liner is scored or scorched, cylinder liner must be replaced.

c. Check cylinder liner projection as follows:

(1) Hold a straight edge along top edge of cylinder liner to be checked.

(2) Using a feeler gage, measure each cylinder liner projection.

(3) Standard projection is 0.0 to 0.004 in. (0.0 to 0.10 mm).

(4) The difference in projection height between any two adjacent cylinders must not exceed 0.0012 in. (0.03 mm).

d. Select piston grade as follows:

(1) The term grade refers to piston diameter and cylinder liner bore combination.

(2) Grade 'A' piston diameter is 3.3804 to 3.3812 in. (85.864 to 85.883 mm) and a Grade 'C' piston diameter is 3.3813 to 3.3820 in. (85.884 to 85.903 mm). These diameters are measured with a micrometer, 2.05 in. (52 mm) from top of piston.

(3) Cylinder liner bore and piston clearance is 0.0057 to 0.0077 in. (0.146 to 0.196 mm).

NOTE

Cylinder liner piston kit clearance are preset, Howerver, cylinder liner
installation procedure may result in slight decrease in cylinder liner bore
clearances. Always measure cylinder liner bore clearance after installa-tion
to be sure that is correct.

e. Measure new cylinder liner bore as follows:

(1) Measure cylinder liner bore at two points, 'A' and 'B', in four different directions (W-W, X-X, Y-Y and Z-Z)
as shown in FIGURE 3-114.

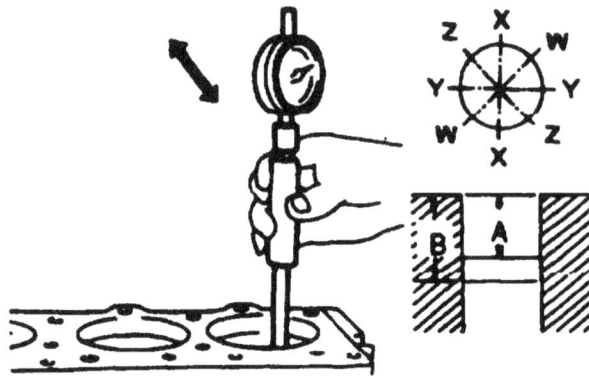

A = 0.79 in. (20 mm)
B = 5.51 in. (140 mm)

FIGURE 3-114. Cylinder Liner Bore Measurement

(2) Calculate average value of the eight measurememts. Standardbore is 3.387 to 3.388 in. (86.021 to 86.060
mm).

(3) Select theappropriate piston grade as described in step d.

NOTE

It is most important that correct piston grade be used. Failure to select correct piston grade will result in engine failure. Always measure cylinder bore and select correct piston grade.

3-32.3. Assembly.

WARNING

Compressed air used for cleaning purposes will not exceed 30 psi (207 kPa).
Use only with effective chip guarding and personal protective equipment (goggles/shield, gloves, etc.).

a. Use compressed air to thoroughly clean cylinder block interior and exterior, oil holes, and water jackets before assembly.

b. Install dowel pins and fittings if removed.

c. Install new cylinder liners (15, FIGURE 3-111) as follows:

WARNING

Diesel fuel is flammable and toxic to eyes, skin, and respiratory tract. Skin/eye protection required. Avoid repeated/prolonged contact. Good general ventilation is normally adequate.

(1) Use clean diesel fuel to thoroughly clean cylinder liners and bores.

(2) Blow dry with compressed air.

NOTE

All foreign material must be carefully removed from cylinder liner and bores before installation.

(3) Insert cylinder liner into cylinder block from top, refer to FIGURE 3-115.

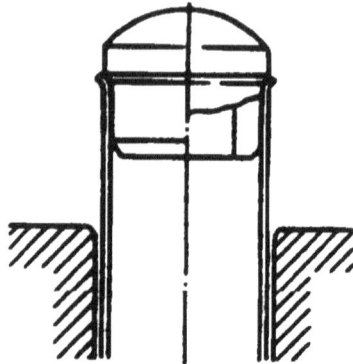

FIGURE 3-115. Cylinder Liner Installation

(4) Set cylinder liner installer (320903) on top of cylinder liner, refer to FIGURE 3-116.

(5) Position cylinder block so installer center is directly beneath arbor press shaft center.

NOTE

Check that cylinder liner is set perpendicular to arbor press and that there is no wobble.

(6) Using arbor press, apply a seating force of 1102.5 lbs (500 kg) to cylinder liner.

(7) Apply a force of 5512.5 lbs (2500 kg) to fully seated cylinder liner.

(8) Measure cylinder liner projection, refer to paragraph 3-32.2, step c.

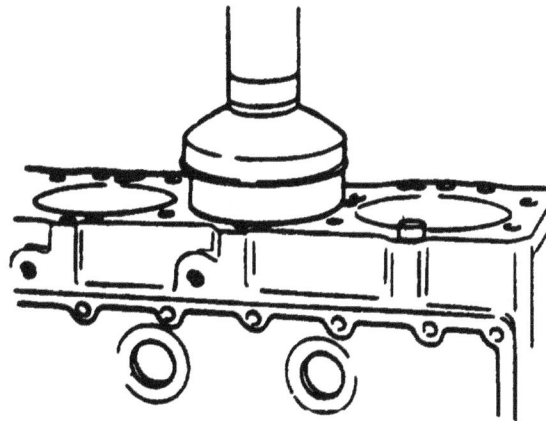

FIGURE 3-116. Pressing Cylinder Liner

d. Positioncover (13, FIGURE 3-111) and new packing (14) on engine block and secure with two screws (12).

e. Install camshaft bearings (11) using bearing remover/ installer (320905) ensuring bearing oil holes align with cylinder block oil holes.

f. Apply a coat of engine lubricating oil (ML-L-21 04) to the eight tappets (10) and cylinder block tappet traveling bores. If reusing tappets, ensure they are installed in correct locations as tagged during disassembly.

g. Apply adhesive (242) to lip of coolant duct (9) and install it in engine block.

h. Apply adhesive (242) to all freeze plugs removed and tap them into holes in block.

i. Install camshaft, refer to paragraph 3-30.5.

j. Install crankshaft, refer to paragraph 3-28.5.

k. Position right engine foot (7) and alternator mounting bracket (6) on cylinder block and secure with bolts (4) and washers (5).

l. Position left engine foot (3) on cylinder block and secure with bolts (1) and washers (2).

m. Install new gasket (40) and front plate (41) on cylinder block aligning front plate with two pins (30). Secure plate with screws (42) and tighten to 168 pounds-inch (19 Nm)

APPENDIX A
REFERENCES

A-1 **SCOPE.**

This appendix lists all forms, field manuals, technical manuals and miscellaneous publications referenced in this manual

A-2 **FORMS.**

Air Force Reporting of Errors Form . . . . AFTO Form 22

Product Quality Deficiency Report . . . . SF 368

Recommended Changes to DA Publications . . . . DA Form 2028-2

Recommended Changes to Publications and Blank Forms . . . . DA Form 2028

Report of Discrepancy (ROD) . . . . SF 384

Reporting of Item and Packaging Discrepancies . . . . AR 735-11-2

Reporting of Transportation Discrepancies in Shipment. . . . . AR 55-38

Transportation Discrepancy Report . . . . SF 361

Equipment Control Records. . . . . DA Form 2408-9

A-3 **FIELD MANUALS.**

First Aid for Soldiers . . . . FM 21-11

A-4 **TECHNICAL MANUALS.**

Destruction of Materiel . . . . TM 750-244-3

A-5 **MISCELLANEOUS PUBLICATIONS.**

Maintenance Management Policy. . . . . AFR 88-1

Preservation of USAMECOM Mechanical
   Equipment for Shipment and Storage. . . . . TB 740-97-2

APPENDIX A
REFERENCES - CONTINUED

A-5 MISCELLANEOUS PUBLICATIONS - Continued

Suggestion Program . . . . AFR 900-4

The Army Maintenance Management System (TAMMS)...................................DA PAM 738-750

USAF Materiel Deficiency Reporting.. TO-OO-35D54

APPENDIX B
MAINTENANCE ALLOCATION CHART
SECTION I.
INTRODUCTION

B-1 General.

a. This section provides a general explanation of all maintenance and repair functions authorized at various maintenance categories.

b. The Maintenance Allocation Chart (MAC) in section II designates overall authority and responsibility for the performance of maintenance functions on the diesel engine and its components. The application of the mainte-nance functions to the engine or components will be consistent with the capacities and capabilities of the desig-nated maintenance categories.

c. Section III lists the tools and test equipment (both special tools and common tool sets) required for each maintenance function as referenced from section II.

d. Section IV contains supplemental instructions and explanatory notes for particular maintenance functions.

B-2 Maintenance Functions. Maintenance functions will be limited to and defined as follows:

a. Inspect. To determine the serviceability of an item by comparing its physical, mechanical, and/or electrical char-acter-istics with established standards through examination (e.g., by sight, sound, or feel).

b. Test. To verify serviceability by measuring the mechanical, pneumatic, or electrical characteristics of an item and comparing those characteristics with prescribed standards.

c. Service. Operations required periodically to keep an item in proper operating condition, i.e., to clean (include decontaminate, when required), to preserve, to drain, to paint, or to replenish fuel, lubricants, chemical fluids, or gases.

d. Adjust. To maintain or regulate, within prescribed limits, by bringing into proper or exact position, or by setting the operating characteristics to specified parameters.

e. Aline. To adjust specified variable elements of an item to bring about optimum or desired performance.

f. Calibrate. To determine and cause corrections to be made or to be adjusted on instruments or Test, Measuring, and Diagnostic Equipment (TMDE) used in precision measurement. Consists of comparisons of two instru-ments, one of which is a certified standardof known accuracy, to detect and adjust any discrepancy in the accu-racy of the instrument being compared.

g. Remove/Install. To remove and install the same item when required to perform service or other maintenance functions. Install may be the act of emplacing, seating, or fixing into position a spare, repair part, or module (component or assembly) in a manner to allow the proper functioning of an equipment or system.

h. Replace. To remove an unserviceable item and install a serviceable counterpart in its place. Replace is autho-rized by the MAC and is shown as the 3rd position code of the SMR code.

i. Repair. The application of maintenance services, including fault location/troubleshooting, removal/installation and disassembly/assembly procedures, and maintenance actions to identify troubles and restore serviceability to an item by correcting specific damage, fault, malfunction, or failure in a part, subassembly, module (com-po-nent or assembly), end item, or system.

j. Overhaul. That maintenance effort (service/action) prescribed to restore an item to a completely serviceable/operational condition as required by maintenance standards in appropriate technical publications (i.e., DMWR). Overhaul is normally the highest degree of maintenance performed by the Army. Overhaul does not normally return an item to like new condition.

k. Rebuild. Consists of those services/actions necessary for the restoration of unserviceable equipment to a like new condition in accordance with original manufacturing standards. Rebuild is the highest degree of material maintenance applied to Army equipment. The rebuild operation includes the act of returning to zero those age measurements (hours/miles, etc.) considered in classifying Army equipment/components.

B-3 Explanation of Columns in the MAC, Section II

a. Column 1, Group Number. Column 1 lists functional group code numbers, the purpose of which are to identify maintenance significant components, assemblies, subassemblies, and modules with the next higher as-sem-bly. End item group number shall be "00".

b. Column 2, Component/Assembly. Column 2 contains the names of components, assemblies, subassemblies, and modules for which maintenance is authorized.

c. Column 3, Maintenance Function. Column 3 lists the function to be performed on the item listed in column 2. (For detailed explanation of these functions, see paragraph B-2).

d. Column 4, Maintenance Category. Column 4 specifies, by the listing of a work time figure in the appropriate s&column(s), the category of maintenance authorized to perform the function listed in column 3. This figure represents the active time required to perform that maintenance function at the indicated category of mainte-nance. If the number or complexity of the tasks within the listed maintenance function vary at different mainte-nance categories, appropriate work time figures will be shown for each category. The work time figure repre-sents the average time required to restore an item (assembly, subassembly, component, module, end item, or system) to a serviceable condition under typical field operating conditions. This time includes preparation time (including any necessary disassembly/assembly time), troubleshooting/fault location time, and quality as-surance/quality control time in addition to the time required to perform the specific tasks identified for the main-tenance functions authorized in the Maintenance Allocation Chart. The symbol designations for the various maintenance categories are as follows:

C . . . . Operator or crew
O . . . . .Organization Maintenance
F . . . . .Direct Support Maintenance
H   General Support Maintenance
D · · · · · Depot Maintenance

e. Column 5, Tools and Equipment. Column 5 specifies, by code, those common tool sets (not individual tools) and special tools, TMDE, and support equipment required to perform the designated function.

f. Column 6, Remarks. This column shall, when applicable, contain a letter code, in alphabetical order, which shall be keyed to the remarks contained in Section IV.

B-4 Explanation of Columns in Tool and Test Equipment Requirements, Section III

a. Column 1, Reference Code. The tool and test equipment reference code correlates with a code used in the MAC, Section II, Column 5.

b. Column 2, Maintenance Category. The lowest category of maintenance authorized to use the tool or test equipment.

c. Column 3, Nomenclature. Name or identification of the tool or test equipment.

    d. Column 4, National Stock Number. The National Stock Number of the tool or test equipment.

    e. Column 5, Tool Number. The manufacturer's part number of the tool or test equipment.

B-5 Explanation of Columns in Remarks, Section IV

    a. Column 1, Reference Code. The code recorded in column 6, II.

    b. Column 2, Remarks. This lists information pretinent to the maintenance function being performed as indicated in the MAC, Section II.

SECTION II. MAINTENANCE ALLOCATION CHART
FOR
DIESEL ENGINE MODEL C-240PW-28

(1) GROUP NUMBER	(2) COMPONENT/ASSEMBLY	(3) MAINTENANCE FUNCTION	(4) MAINTENANCE CATEGORY					(5) TOOLS AND EQUIP.	(6) REMARKS
			C	O	F	H	D		
00	ENGINE ASSEMBLY 15KW	INSPECT SERVICE TEST ADJUST REPAIR R/I REPLACE OVERHAUL							
0100	COOLING SYSTEM	INSPECT SERVICE REPAIR							
0101	THERMOSTAT	INSPECT REPLACE							
0102	WATER PUMP	INSPECT R/I REPLACE							
0200	FUEL SYSTEM	INSPECT SERVICE REPAIR							
0201	FUEL FILTER	INSPECT							
0202	FUEL INJECTOR	INSPECT ADJUST TEST R/I REPLACE REPAIR							
0203	FUEL INJECTION PUMP	INSPECT TEST R/I REPLACE REPAIR							
0300	LUBRICATION SYSTEM	INSPECT SERVICE REPAIR							
0301	OIL FILTER	INSPECT REPLACE							
0302	OIL PRESSURE RELIEF VALVE	INSPECT TEST R/I REPLACE							
0303	OIL PAN	INSPECT REPLACE							
0304	OIL PUMP ASSEMBLY	INSPECT R/I REPLACE REPAIR							
0400	INTAKE/EXHAUST SYSTEM	INSPECT REPAIR							

SECTION II. MAINTENANCE ALLOCATIONCHART - Continued
FOR
DIESEL ENGINE MODEL C-240PW-28

(1) Group Number	(2) Component/Assembly	(3) Maintenance Function	(4) Maintenance Category					(5) Tools and Equip.	(6) Remarks
			C	O	F	H	D		
0401	PCV ASSEMBLY	INSPECT		.2					
		SERVICE		.5				5	
		R/I		.5				5	
		REPLACE		.5				5	C
		REPAIR		.7				5	
0402	INTAKE MANIFOLD	INSPECT		.1					
		R/I		1.0				1, 5	
		REPLACE		1.0				1, 5	C
0403	EXHAUST MANIFOLD	INSPECT		.1					
		R/I		1.0				1, 5	
		REPLACE		1.0				1, 5	C
0500	ELECTRICAL SYSTEM	INSPECT	.1	.1					A
		REPAIR		1.0	1.0				
0501	GLOW PLUGS	INSPECT		.1					
		TEST		.5				1	
		R/I		1.0				1, 5	
		REPLACE		1.0				1, 5	C
0502	STARTER MOTOR ASSY	INSPECT		.1					
		TEST		.5	1.0			1	
		R/I		1.0				5	
		REPLACE		1.0				5	C
		REPAIR		1.5	2.0			1, 5	
0503	ALTERNATOR ASSEMBLY	INSPECT		.1					
		TEST		.5	1.0			1	
		R/I		1.0				5	
		REPLACE		1.0				5	C
		REPAIR			2.0			5	
	ENGINE BLOCK ASSEMBLY	INSPECT	.1	.5	.5	.5			
		ADJUST			1.0				
		TEST			1.0	1.0			
		REPAIR		.5	2.5	3.0			
0601	FRONT GEAR COVER	INSPECT			.1				
		R/I			3.0			4, 5	
		REPLACE			3.0			4, 5	C
		REPAIR			3.5				
060101	SEAL, CRANKSHAFT WONT	INSPECT			.1				
		ADJUST			3.5			5, 16	
0602	ROCKER ARM ASSEMBLY	TEST			.5			5	
		R/I			1.0			3, 4	
		REPLACE			1.0			4, 5	
		REPAIR			1.0			4, 5	C
		INSPECT			1.0			4, 5	
		TEST			1.5				
0603	CYLINDER HEAD ASSY	R/I			.2				
		REPLACE			1.0			3, 4, 5	
		REPAIR			4.0			1, 5	
		INSPECT			4.0			1, 5	C
		TEST			6.0			3, 4, 5, 17	
0604	VALVE SPRINGS	R/I			.3				
		REPLACE			.5			3	
					5.0			3, 5	
					5.0			3, 5	C

SECTION II. MAINTENANCE ALLOCATION CHART - Continued
FOR
DIESEL ENGINE MODEL C-240PW-28

(1) GROUP NUMBER	(2) COMPONENT/ASSEMBLY	MAINTENANCE FUNCTION	(4) MAINTENANCE CATEGORY					TOOLS EQUIP.	(6) REMARKS
			C	O	F	H	D		
0605	INTAKE/EXHAUST VALVES	INSPECT			.5			3	
		TEST			5.0			3,5	
		R/I						3.5	
		REPLACE			5.0				C
		REPAIR			6:0			3	
0606	CONNECTING ROD & PISTON ASSEMBLY	INSPECT				.5			
		TEST				1.0		3,4.5	
		R/I				8.0			
		REPLACE				8.0		1,3,5	C
		REPAIR				9.0		1,4,5	
0607	IDLER GEAR ASSEMBLY	INSPECT			.2				
		TEST			1.0			3,5	
		R/I			3.5			1.5	
		REPLACE			3.5			1,5	C
		REPAIR			4.5			1.5	
0606	CAMSHAFT ASSEMBLY	INSPECT				.5			
		TEST				1.0		3,5	
		R/I				4.0		5	
		REPLACE				4.0		5	C
		REPAIR				5.0		1,4,5	
0609	FLYWHEEL ASSEMBLY	INSPECT				.2			
		R/I				1.0			
		REPLACE				1.0		1,5	C
		REPAIR				2.0		1,5	
0610	SEAL, CRANKSHAFT REAR	INSPECT				.1			
		REPLACE				1.0		5, 12	
0611	CRANKSHAFT ASSEMBLY	INSPECT				.5			
		TEST				1.0		3	
		R/I				10.0		1,5	
		REPLACE				10.0		1,5	C
		REPAIR				11.0		3,4, 5, 10	
0612	MAIN & ROD BEARINGS	INSPECT							
		TEST				1.5		3	
						11.5		1.5	
		REPLACE				11.5		1,5	C
0613	CRANKCASE ASSEMBLY	INSPECT				.5			
		TEST				1.0			
						12.0		3,4,5	
		REPLACE				12.0		1,5	C
		REPAIR				14.0			
061301	CYLINDER LINERS	INSPECT				2			
		TEST				1.0			
		REPLACE				2.0		5,11,14,15	
061302	CAMSHAFT BEARINGS	INSPECT				.2			
		TEST				1.0		3	
		REPLACE				2.0		5, 13	

SECTION III. TOOL AND TEST EQUIPMENT REQUIREMENTS
FOR
DIESEL ENGINE MODEL C-240PW-28

TOOL OR TEST EQUIPMENT REF CODE	MAINTENANCE CATEGORY	NOMENCLATURE	NATIONAL/STOCK NUMBER	TOOL NUMBER
1	O, F, H	SHOP EQUIPMENT. AUTOMOTIVE MAINTE-NANCE AND REPAIR: ORGANIZATIONAL MAIN-TENANCE COMMON NO. 1, LESS POWER	4910-00-754-0654	SC4910-95 CL-A74
2	F	SHOP EQUIPMENT. FUEL AND ELECTRICAL SYSTEM ENGINE. FIELD MAINTENANCE BASIC, LESS POWER	4940-00-754-0714	SC4910-95 CL-820
3	H	SHOP EQUIPMENT, AUTOMOTIVE MAINTE-NANCE AND REPAIR: FIELD MAINTENANCE, SUPPLEMENTAL SET NO. 2, LESS POWER	4910-00-754-0707	SC4910-95 CL-A63
4	F,H	TOOL SET, BASIC, FIELD MAINTENANCE	4910-00-754-0705	SC4910-95-CL A31
5	O, F, H	TOOL KIT, GENERAL MECHANICS	5160-00-177-7033	SC5180-90-CL-N26
6	F	TESTER, PRESSURE GAGE	6685-00-263-0658	MIL-T-2764
7	F	ADAPTER, COMPRESSION GAGE	4910-01-196-8670	J-26999-2
6	F, H	TEST SET, FUEL INJECTION PUMP	4910-01-121-6869	77-7028
9	H	PLUNGER, SPRING HOLDER	-	157931-4100
10	H	INSTALLER, CRANKSHAFT GEAR		320906
11	H	GRIP	5120-01-263-8752	7631
12	F	INSTALLER, CRANKSHAFT REAR OIL SEAL	5340-01-265-6734	NU7627
13	H	REMOVER/INSTALLER, CAMSHAFT BEARING	5120-01-268-8826	320905
14	H	INSTALLER. CYLINDER LINER	5120-01-263-8751	320903
15	H	REMOVER,CYLINDERLINER	5120-01-263-8749	NU-7628
16	H	INSTALLER, OIL SEAL	5120-01-263-8750	5-8522-0013-0
17	F	REPLACER, VALVE GUIDE	5120-01-268-8822	NW7634
16	H	SPANNER	5120-01-266-6823	157916-5320
19	H	EXTRACTOR	-	157926-6220
20	H	BRACKET	-	157944-7620
21	H	UNIVERSAL VISE	-	157944-6520
22	H	COUPUNG	-	157842-4420!57916-5420
23	H	SPANNER	-	157921-0120
24	H	TAPPET INSTALLER	-	157931-6120
25	H	CLAMP	5120-01-056-1997	157920-9620
26	H	EXTRACTOR	-	157924-1620
27	H	INSERTER TOOL	-	105782-6280
26	H	MEASURING DEVICE	-	106782-4020
29	H	MEASURING DEVICE	-	

SECTION IV. REMARKS
FOR
DIESEL ENGINE MODEL C-240PW-28

REFERENCE CODE	REMARKS
A	Refer to end item operator's manual.
B	Refer to end item maintenance manual.
C	Replace function is identical to removal/install function.
D	Fuel filter does not include a filter element for this application. Refer to end item maintenancemanual for fuel filter requirement.
E	Repair limited to replacement of gear.

APPENDIX C
EXPENDABLE/DURABLE SUPPLIES AND MATERIALS LIST
SECTION I. INTRODUCTION

C-1 **SCOPE.**

This appendix lists expendable supplies and materials you will need to operate and maintain the engine: These items are authorized to you by CTA 50-970, Expendable Items (except Medical, Class V, Repair Parts, and Heraldic Items).

C-2 **EXPLANATION OF COLUMNS.**

a. Column (1) - Item Number. This number is assigned to each entry in the listing.

b. Column (2) - Level. This column identifies the lowest level of maintenance that requires the listed item.

c. Column (3) - National Stock Number. This is the National Stock Number assigned to the item; use it to request or requisition the item.

d. Column (4) - Description. Indicates the Federal item name and, if required, a description to identify the item. The last line for each item indicates the Commercial and Government Entity (CAGE) code in parentheses ( ) followed by the part number.

e. Column (5) - Unit of Measure (U/M). Indicates the measure used in performing the actual maintenance function.

    This measure is expressed by a two-character alphabetical abbreviation (e.g., ea, in, pr). If the unit of measures differs from the unit of issue, requisition the lowest unit of issue that will satisfy your requirements.

APPENDIX C
EXPENDABLE/DURABLE SUPPLIES AND MATERIALS LIST - CONTINUED

(1) ITEM NUMBER	(2) LEVEL	(3) NATIONAL STOCK NUMBER	(4) DESCRIPTION	(5) U/M
1	DS/GS	8040-01-250-3969	Adhesive, LOCTITE 242	Tu
2	DS/GS	8040-00-390-7959	Adhesive, Seal, 847	Tu
3	DS/GS	8040-01-055-6407	Adhesive, Silicon 584R46001101	TU
4	DS/GS	6850-00-311-6172	Bearing Dye, Blue, HISPOTBLUE 107	EA
5	Units/ DS/GS	7920-01-338-3329	Cloth, Cleaning, TX- 1250	EA
6	DS/GS	5350-00-009-1751	Compound, Lapping, SSL1682	OZ
7	DS/GS	5210-00-640-6178	Gage, Bearing Clear, PLASTIGAGEPR1	BX
8	DS/GS	9150-00-053-0101	Grease Molybdenum Disulfide MOLYKOTEZ	TU
9	Unit/ DS/GS	9150-00-152-4117	lubricating Oil, Eng 15/40W, MIL-L-2104	QT
10	Unit/ DSGS	5350-00-224-7201	Paper, Abrasive, #400	SHT
11	Unit./ DS/GS	8030-00-849-0071	Sealing Compound, FORMAGASKET2	TU
12	DS/GS	3439-00-974-1873	Solder, Tn Alloy, SN60WRAP2,1 lb Spool	OZ
13	Unit/ DS/GS	6850-00-264-9038	Solvent, Dry Cleaning, P-D-680,5 Gal. Can	GL

APPENDIX D
FABRICATION OF TOOLS

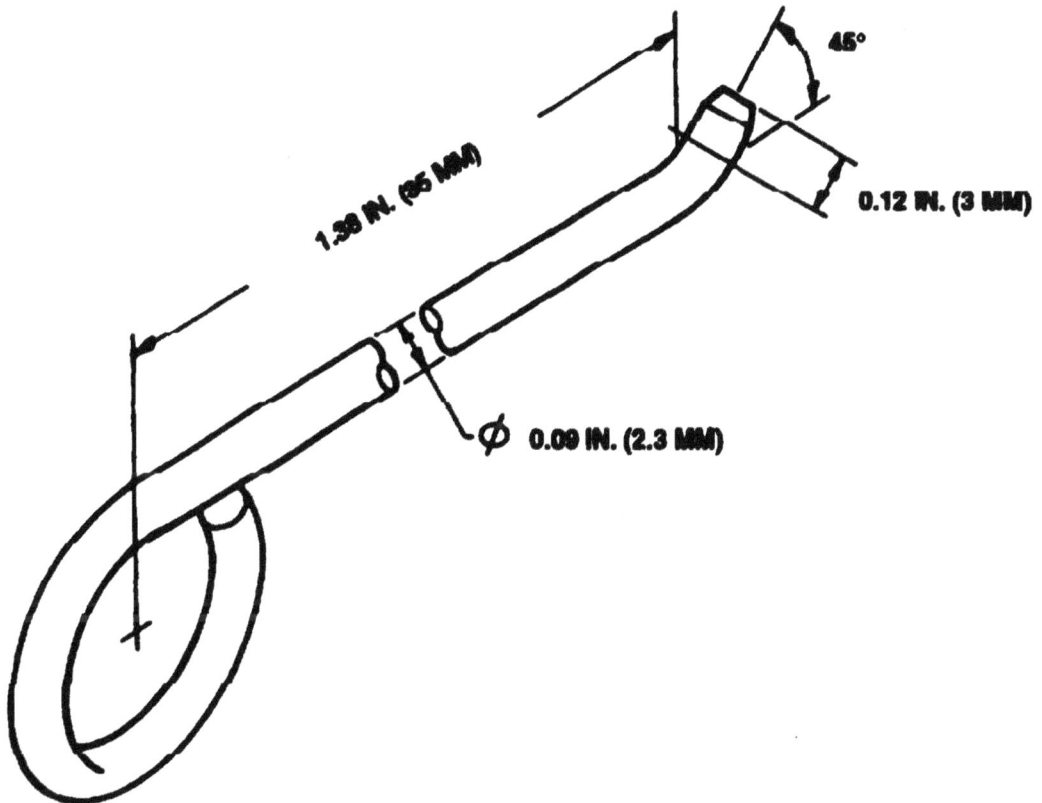

FIGURE D-1. Fuel Injection Pump Tappet Holder

Appendix E. MAINTENANCE PROCEDURE AND AU-

Section I. INTRODUCTION

SCOPE. This appendix shall be used when a Commercial Off The Shelf (COTS) manual is used by the Military services. This appendix is divided into three sections. Section I gives a brief description of the sections within this appendix. Section II list the paragraphs to the associated maintenance level. Section III list the maintenance level to the associated paragraphs.

Section II. PARAGRAPH TO MAINTENANCE LEVEL

Paragraph	Maintenance Level
3-4	O
3-5	O
3-6	O
3-7	O
3-8	O
3-9	O, F
3-10	F, H
3-11	F, H
3-12	F
3-13	O
3-14	F
3-15	F
3-16	F
3-17	F
3-18	O
3-19	O
3-20	O, F
3-21	O, F
3-22	F
3-23	F
3-24	F
3-25	H
3-26	F
3-27	H
3-28	F
3-29	H
3-30	H

Section 111. MAINTENANCE LEVELTD PARAGRAPHS

Maintenance Level	Paragraphs
ALL	3-4
O	3-5, 3-6, 3-7, 3-8, 3-9, 3-13, 3-17, 3-18, 3-19, 3-20, 3-21
F	3-9, 3-10, 3-11, 3-12, 3-14, 3-16, 3-20, 3-21, 3-22, 3-23, 3-24, 3-26, 3-28
H	3-10, 3-11, 3-25, 3-27, 3-29, 3-30

INDEX

INDEX-Continued

INDEX - Continued

<u>Description</u>   <u>Page No.</u>

T

V

W

By Order of the Secretaries of the Army and Air Force:

GORDON R. SULLIVAN
General, United States Army
Chief of Staff

Official:

MILTON H. HAMILTON
Administrative Assistant to the
Secretary of the Army

MERRILL A. McPeak
General USAF
Chief of Staff

Official:

RONALD W. YATES
General, USAF
Commander, Air Force Material Command

DISTRIBUTION:
    To be distributed in accordance with DA Form 12-25-E. block no. 5143, requirements for
TM 9-2815-254-24.

*U. S. GOVERNMENT PRINTING OFFICE: O - 432-785 (81012)

These are the instructions for sending an electronic 2028

The following format must be used if submitting an electronic 2028. The subject line must be exactly the same and all fields must be included; however only the following fields are mandatory: 1, 3, 4, 5, 6, 7, 8, 9, 10, 13, 15, 16, 17, and 27.

From: "Whomever"<whomever@avma27.army.mil> To: mp-mt%avma28@st-louis-emh7.army.mil

Subject: DA Form 2028

1. **From:** Joe Smith
2. Unit: home
3. *Address:* 4300 Park
4. *City:* Hometown
5. *St:* MO
6. *Zip:* 77777
7. *Date Sent:* 19-OCT-93
8. *Pub no:* 55-2840-229-23
9. *Pub Title:* TM
10.   *Publication Date:* 04-JUL-85
11.   Change *Number:* 7
12.   *Submitter Rank: MSG*
13.   ***Submitter FName:*** Joe
14.   *Submitter MName:* T
15.   ***Submitter LName:*** Smith
16.   ***Submitter Phone:*** 123-123-1234
17.   ***Problem:* 1**
18.   Page: 2
19.   *Paragraph: 3*
20.   *Line: 4*
21.   NSN: 5
22.   *Reference: 6*
23.   *Figure: 7*
24.   *Table: 8*
25.   *Item: 9*
26.   *Total:* 123
27.   **Text:**
This is the text for the problem below line 27.

The Metric System and Equivalents

Linear Measure

1 centimeter = 10 millimeters = .39 inch
1 decimeter = 10 centimeters = 3.94 inches
1 meter = 10 decimeters = 39.37 inches
1 dekameter = 10 meters = 32.8 feet
1 hectometer = 10 dekameters = 328.08 feet
1 kilometer = 10 hectometers = 3,280.8 feet

Weights

1 centigram = 10 milligrams = .15 grain
1 decigram = 10 centigrams = 1.54 grains
1 gram = 10 decigram = .035 ounce
1 dekagram = 10 grams = .35 ounce
1 hectogram = 10 dekagrams = 3.52 ounces
1 kilogram = 10 hectograms = 2.2 pounds
1 quintal = 100 kilograms = 220.46 pounds
1 metric ton = 10 quintals = 1.1 short tons

Liquid Measure

1 centiliter = 10 milliliters = .34 fl. ounce
1 deciliter = 10 centiliters = 3.38 fl. ounces
1 liter = 10 deciliters = 33.81 fl. ounces
1 dekaliter = 10 liters = 2.64 gallons
1 hectoliter = 10 dekaliters = 26.42 gallons
1 kiloliter = 10 hectoliters = 264.18 gallons

Square Measure

1 sq. centimeter = 100 sq. millimeters = .155 sq. inch
1 sq. decimeter = 100 sq. centimeters = 15.5 sq. inches
1 sq. meter (centare) = 100 sq. decimeters = 10.76 sq. feet
1 sq. dekameter (are) = 100 sq. meters = 1,076.4 sq. feet
1 sq. hectometer (hectare) = 100 sq. dekameters = 2.47 acres
1 sq. kilometer = 100 sq. hectometers = .386 sq. mile

Cubic Measure

1 cu. centimeter = 1000 cu. millimeters = .06 cu. inch
1 cu. decimeter = 1000 cu. centimeters = 61.02 cu. inches
1 cu. meter = 1000 cu. decimeters = 35.31 cu. feet

Approximate Conversion Factors

To change	To	Multiply by	To change	To	Multiply by
inches	centimeters	2.540	ounce-inches	newton-meters	.007062
feet	meters	.305	centimeters	inches	.394
yards	meters	.914	meters	feet	3.280
miles	kilometers	1.609	meters	yards	1.094
square inches	square centimeters	6.451	kilometers	miles	.621
square feet	square meters	.093	square centimeters	square inches	.155
square yards	square meters	.836	square meters	square feet	10.764
square miles	square kilometers	2.590	square meters	square yards	1.196
acres	square hectometers	.405	square kilometers	square miles	.386
cubic feet	cubic meters	.028	square hectometers	acres	2.471
cubic yards	cubic meters	.765	cubic meters	cubic feet	35.315
fluid ounces	milliliters	29,573	cubic meters	cubic yards	1.308
pints	liters	.473	milliliters	fluid ounces	.034
quarts	liters	.946	liters	pints	2.113
gallons	liters	3.785	liters	quarts	1.057
ounces	grams	28.349	liters	gallons	.264
pounds	kilograms	.454	grams	ounces	.035
short tons	metric tons	.907	kilograms	pounds	2.205
pound-feet	newton-meters	1.356	metric tons	short tons	1.102
pound-inches	newton-meters	.11296			

Temperature (Exact)

°F	Fahrenheit temperature	5/9 (after subtracting 32)	Celsius temperature	°C

*ARMY TM 9-2815-254-24P
AIR FORCE TO 38G1-94-2

TECHNICAL MANUAL

FOR

FIELD AND SUSTAINMENT

REPAIR PARTS AND SPECIAL TOOLS LIST

DIESEL ENGINE, FOUR-CYLINDER, FOUR CYCLE, FUEL INJECTED C-240PW-28

(NSN: 2815-01-350-2207)

HEADQUARTERS, DEPARTMENTS OF THE ARMY AND AIR FORCE
15 JUNE 2010

LIST OF EFFECTIVE PAGES / WORK PACKAGES

NOTE: This manual supersedes TM 9-2815-254-24P and TO 38G1-94-4 dated 30 October 1996. Date of issue for the revised manual is: 15 June 2010. Zero in the "Change No." column indicates an original page or work package.

Date of issue for revision is:

Change 15 JUNE 2010

TOTAL NUMBER OF PAGES FOR FRONT AND REAR MATTER IS 22 AND TOTAL NUMBER OF WORK PACKAGES IS 24, CONSISTING OF THE FOLLOWING:

Page / WP No.	Change No.	Page / WP No.	Change No.
Front Cover		0	
Blank		0	
i -iii		0	
iv Blank		0	
Chp 1 title page		0	
Chp 1 Index		0	
WP 0001 (8 pgs)		0	
WP 0002 (4 pgs)		0	
WP 0003 (4 pgs)		0	
WP 0004 (4 pgs)		0	
WP 0005 (6 pgs)		0	
WP 0006 (4 pgs)		0	
WP 0007 (4 pgs)		0	
WP 0008 (4 pgs)		0	
WP 0009 (4 pgs)		0	
WP 0010 (4 pgs)		0	
WP 0011 (4 pgs)		0	
WP 0012 (4 pgs)		0	
WP 0013 (4 pgs)		0	
WP 0014 (4 pgs)		0	
WP 0015 (4 pgs)		0	
WP 0016 (4 pgs)		0	
WP 0017 (4 pgs)		0	
WP 0018 (4 pgs)		0	
WP 0019 (4 pgs)		0	
WP 0020 (6 pgs)		0	
WP 0021 (2 pgs)		0	
WP 0022 (6 pgs)		0	
WP 0023 (8 pgs)		0	
WP 0024 (10 pgs)		0	
Inside back cover		0	
Back cover		0	

HEADQUARTERS,
DEPARTMENTS OF THE ARMY AND AIR FORCE
WASHINGTON, D.C., 15 JUNE 2010

TECHNICAL MANUAL
TECHNICAL MANUAL
FIELD AND SUSTAINMENT

REPAIR PARTS AND SPECIAL

DIESEL ENGINE, FOUR-CYLINDER,

FOUR CYCLE, FUEL INJECTED

C-240PW-28

(NSN: 2815-01-350-2207) (EIC: N/A)

(NSN: 2815-01-350-2207)

REPORTING ERRORS AND RECOMMENDING IMPROVEMENTS

You can help improve this manual. If you find any mistakes or if you know of a way to improve the procedures, please let us know. Reports, as applicable by the requiring Service, should be submitted as follows:

(a) (A) Army - Mail your letter or DA Form 2028 (Recommended Changes to Publications and Blank Forms), located in the back of this manual, directly to: Commander, U.S. Army CECOM (LCMC) and Fort Monmouth, ATTN: AMSEL-LC-LEO-E-CM, Fort Monmouth, NJ 07703-5006. You may also send in your recommended changes via electronic mail or by fax. Our fax number is 732-532-3421, DSN 992-3421. Our e-mail address is MONM-AMSELLEOPUBSCHG@conus.army.mil. Our online web address for entering and submitting DA Form 2028s is http://edm.monmouth.army.mil/pubs/2028.html.

(d) (F) Air Force - By Air Force AFTO Form 22 (Technical Manual (TM) change Recommendation and Reply) in accordance with paragraph 6-5, Section VI, TO 00 5 1 directly to prime ALC/MST. A
reply will be furnished to you.

TABLE OF CONTENTS

TABLE OF CONTENTS - Continued

CHAPTER 1

FIELD AND SUSTAINMENT REPAIR PARTS AND SPECIAL TOOLS LIST

FOR

DIESEL ENGINE, FOUR-CYLINDER, FOUR CYCLE, FUEL INJECTED
C-240PW-28
(NSN: 2815-01-350-2207)

CHAPTER 1

PARTS INFORMATION

WORK PACKAGE INDEX

FIELD AND SUSTAINMENT MAINTENANCE

DIESEL ENGINE, FOUR-CYLINDER, FOUR CYCLE, FUEL INJECTED C-240PW-28
REPAIR PARTS AND SPECIAL TOOLS LIST (RPSTL) INTRODUCTION

INTRODUCTION

SCOPE

This RPSTL lists and authorizes spares and repair parts; special tools; special test, measurement, and diagnostic equipment (TMDE); and other special support equipment required for performance of field and sustainment maintenance of the Diesel Engine, Model C-240PW-28. It authorizes the requisitioning, issue, and disposition of spares, repair parts, and special tools as indicated by the source, maintenance, and recoverability (SMR) codes.

GENERAL

In addition to the Introduction work package, this RPSTL is divided into the following work packages.

1. Repair Parts List Work Packages. Work packages containing lists of spares and repair parts authorized by this RPSTL for use in the performance of maintenance. These work packages also include parts which must be removed for replacement of the authorized parts. Parts lists are composed of functional groups in ascending alphanumeric sequence, with the parts in each group listed in ascending figure and item number sequence. Sending units, brackets, filters, and bolts are listed with the component they mount on. Bulk materials are listed by item name in FIG. BULK at the end of the work packages. Repair parts kits are listed separately in their own functional group and work package Repair parts for reparable special tools are also listed in a separate work package. Items listed are shown on the associated illustrations.

2. Special Tools List Work Packages. Work packages containing lists of special tools, special TMDE, and special support equipment authorized by this RPSTL (as indicated by Basis of Issue (BOI) information in the DESCRIPTION AND USABLE ON CODE (UOC) column). Tools that are components of common tool sets and/or Class VII are not listed.

3. Cross-Reference Indexes Work Packages. There are 2 Cross-Reference indexes work packages in this RPSTL: the National Stock Number (NSN) Index work package, and the Part Number (P/N) Index work package. The National Stock Number Index work package refers you to the figure and item number. The Part Number Index work package refers you to the figure and item number.

EXPLANATION OF COLUMNS IN THE REPAIR PARTS LIST AND SPECIAL TOOLS LIST WORK PACKAGES

ITEM NO. (Column (1)). Indicates the number used to identify items called out in the illustration.

SMR CODE (Column (2)). The SMR code containing supply/requisitioning information, maintenance level authorization criteria, and disposition instruction, as shown in the following breakout. This entry may be subdivided into 4 subentries, one for each service.

Table 1. SMR Code Explanation.

<u>Source Code Maintenance Code Recoverability Code</u>

XX   XX   X 1st two positions: 3rd position: 4th position: 5th

position:

How to get an item. Who can install, replace, or use the item.   Who can do complete   repair* on the item. Who determines disposition action on unser-

*Complete Repair: Maintenance capacity, capability, and authority to perform all corrective maintenance tasks of the "Repair" function in a use/user environment in order to restore serviceability to a failed item.

Source Code. The source code tells you how you get an item needed for maintenance, repair, or overhaul of an end item/equipment. Explanations of source codes follow:

Source Code   **Application/Explanation**

PA

PB   **NOTE**

PC   Items coded PC are subject to deterioration.

PD   Stock items; use the applicable NSN to requisition/
                                                       request items with these source codes. They are
PE   authorized to the level indicated by the code entered in
                                                       the third position of the SMR code.
PF

PG

PH

PR

PZ

KD Items with these codes are not to be requested/

KF   requisitioned individually. They are part of a kit which is
                                                       authorized to the maintenance level indicated in the third
KB   position of the SMR code. The complete kit must be
                                                       requisitioned and applied.

MF-Made at field Items with these codes are not to be requisitioned/

MH-Made at below depot/sustainment level   requested individually. They must be made from bulk
                                                       material which is identified by the part number in the
ML-Made at SRA   DESCRIPTION AND USABLE ON CODE (UOC) col-
                                                       umn and listed in the bulk material group work package
MD-Made at depot   of the RPSTL. If the item is authorized to you by the
                                                       third position code of the SMR code, but the source
MG-Navy only   code indicates it is made at higher level, order the item
                                                       from the higher level of maintenance.

AF-Assembled by field Items with these codes are not to be requested/

AH-Assembled by below depot sustainment level   requisitioned individually. The parts that make up the
                                                       assembled item must be requisitioned or fabricated and
AL-Assembled by SRA   assembled at the level of maintenance indicated by
                                                       the source code. If the third position of the SMR code
AD-Assembled by depot   authorizes you to replace the item, but the source code
                                                       indicates the item is assembled at a higher level, order
AG-Navy only   the item from the higher level of maintenance.

XA Do not requisition an "XA" coded item. Order the next
                                                       higher assembly. (Refer to NOTE below.)

XB If an item is not available from salvage, order it using
                                                       the CAGEC and part number.

XC Installation drawings, diagrams, instruction sheets, field
                                                       service drawings; identified by manufacturer's part
                                                       number.

XD Item is not stocked. Order an XD-coded item through
                                                       local purchase or normal supply channels using the
                                                       CAGEC and part number given, if no NSN is available.

NOTE

Cannibalization or controlled exchange, when authorized, may be used as a source of supply for items with the above source codes except for those items source coded "XA" or those aircraft support items restricted by requirements of AR 750-1.

Maintenance Code. Maintenance codes tell you the level(s) of maintenance authorized to use and repair support items. The maintenance codes are entered in the third and fourth positions of the SMR code as follows:

Third Position. The maintenance code entered in the third position tells you the lowest maintenance level authorized to remove, replace, and use an item. The maintenance code entered in the third position will indicate authorization to the following levels of maintenance:

Maintenance Code Application/Explanation

F - Field maintenance can remove, replace, and use the item.

H - Below Depot Sustainment maintenance can remove, replace, and use the item.

L - Specialized repair activity can remove, replace, and use the item.

G - Afloat and ashore intermediate maintenance can remove, replace, and use the item (Navy
          only)

K - Contractor facility can remove, replace, and use the item.

Z - Item is not authorized to be removed, replaced, or used at any maintenance level D - Depot can remove,

replace, and use the item.

*NOTE - Army may use C in the third position. However, for joint service publications, Army will use O.

Fourth Position. The maintenance code entered in the fourth position tells you whether or not the item is to be repaired and identifies the lowest maintenance level with the capability to do complete repair (perform all authorized repair functions).

NOTE

Some limited repair may be done on the item at a lower level of maintenance, if authorized by the Maintenance Allocation Chart (MAC) and SMR codes.

Maintenance Code Application/Explanation

F - Field is the lowest level that can do complete repair of the item.

H - Below Depot Sustainment is the lowest level that can do complete repair of the item.

L - Specialized repair activity is the lowest level that can do complete repair of the item.

D - Depot is the lowest level that can do complete repair of the item.

G - Both afloat and ashore intermediate levels are capable of complete repair of item. (Navy
          only)

K - Complete repair is done at contractor facility

Z - Nonreparable. No repair is authorized.

B - No repair is authorized. No parts or special tools are authorized for maintenance of "B" coded item. However, the item may be reconditioned by adjusting, lubricating, etc., at the user level.

Recoverability Code. Recoverability codes are assigned to items to indicate the disposition action on

**Recoverability
Code**　　　　**Application/Explanation**

Z - Nonreparable item. When unserviceable, condemn and dispose of the item at the level of maintenance shown in the third position of the SMR code.

F - Reparable item. When uneconomically reparable, condemn and dispose of the item at the field level.

H - Reparable item. When uneconomically reparable, condemn and dispose of the item at the below depot sustainment level.

D - Reparable item. When beyond lower level repair capability, return to depot. Condemnation and disposal of item are not authorized below depot level.

L - Reparable item. Condemnation and disposal not authorized below Specialized Repair Activity (SRA).

A - Item requires special handling or condemnation procedures because of specific reasons (such as precious metal content, high dollar value, critical material, or hazardous material). Refer to appropriate manuals/directives for specific instructions.

G - Field level reparable item. Condemn and dispose at either afloat or ashore intermediate levels. (Navy only)

K - Reparable item. Condemnation and disposal to be performed at contractor facility.

NSN (Column (3)). The NSN for the item is listed in this column.

CAGEC (Column (4)). The Commercial and Government Entity Code (CAGEC) is a five-digit code which is used to identify the manufacturer, distributor, or Government agency/activity that supplies the item.

PART NUMBER (Column (5)). Indicates the primary number used by the manufacturer (individual, company, firm, corporation, or Government activity), which controls the design and characteristics of the item by means of its engineering drawings, specifications, standards, and inspection requirements to identify an item or range of items.

NOTE

When you use an NSN to requisition an item, the item you receive may have a different part number from the number listed.

DESCRIPTION AND USABLE ON CODE (UOC) (Column (6)). This column includes the following information:

1. The federal item name, and when required, a minimum description to identify the item.
2. Part numbers of bulk materials are referenced in this column in the line entry to be manufactured or fabricated.
3. Hardness Critical Item (HCI). A support item that provides the equipment with special protection from electromagnetic pulse (EMP) damage during a nuclear attack.
4. The statement END OF FIGURE appears just below the last item description in column (6) for a given figure in both the repair parts list and special tools list work packages.

QTY (Column (7)). The QTY (quantity per figure) column indicates the quantity of the item used in the breakout shown on the illustration/figure, which is prepared for a functional group, subfunctional group, or an assembly. A "V" appearing in this column instead of a quantity indicates that the quantity is variable and quantity may change from application to application.

EXPLANATION OF CROSS-REFERENCE INDEXES WORK PACKAGES FORMAT AND COLUMNS

1. National Stock Number (NSN) Index Work Package. NSN's in this index are listed in National Item Identification Number (NIIN) sequence.

> STOCK NUMBER Column. This column lists the NSN in NIIN sequence. The NIIN consists of the last nine digits of the NSN. When using this column to locate an item, ignore the first four digits of the NSN. However, the complete NSN should be used when ordering items by stock number.

> For example, if the NSN is 5385-01-574-1476, the NIIN is 01-574-1476.

> FIG. Column. This column lists the number of the figure where the item is identified/located. The figures are in numerical order in the repair parts list and special tools list work packages.

> ITEM Column. The item number identifies the item associated with the figure listed in the adjacent FIG. column. This item is also identified by the NSN listed on the same line.

2. Part Number (P/N) Index Work Package. Part numbers in this index are listed in ascending alphanumeric sequence (vertical arrangement of letter and number combinations which places the first letter or digit of each group in order A through Z, followed by the numbers 0 through 9 and each following letter or digit in like order).

> PART NUMBER Column. Indicates the part number assigned to the item.

> FIG. Column. This column lists the number of the figure where the item is identified/located in the repair parts list and special tools list work packages.

> ITEM Column. The item number is the number assigned to the item as it appears in the figure referenced in the adjacent figure number column.

SPECIAL INFORMATION

UOC. The UOC appears in the lower left corner of the Description Column heading. Usable on codes are shown as "UOC: ..." in the Description Column (justified left) on the first line under the applicable item/nomenclature. Uncoded items are applicable to all models. Identification of the UOCs used in the RPSTL are:

Code Used On

ERC Diesel Engine, Model C-240PW-28

Fabrication Instructions. Bulk materials required to manufacture items are listed in the bulk material functional group of this RPSTL. Part numbers for bulk material are also referenced in the Description Column of the line item entry for the item to be manufactured/fabricated. Detailed fabrication instructions for items source coded to be manufactured or fabricated are found in TM 9-2815-254-24.

Index Numbers. Items which have the word BULK in the figure column will have an index number shown in the item number column. This index number is a cross-reference between the NSN / Part Number (P/N) Index work packages and the bulk material list in the repair parts list work package.

Associated Publications. The publication(s) listed below pertains to the Diesel Engine, Model C-240PW-28:

Publication Short Title

TM 9-2815-254-24 Diesel Engine, Four-cylinder, Four Cycle, Fuel Injected - C-240PW-28

Illustrations List. The illustrations in this RPSTL contain field authorized items. Illustrations published in (enter applicable TM number for the higher maintenance level RPSTL, e.g., for field, below depot sustainment, etc.) that contain field authorized items also appear in this RPSTL. The tabular list in the repair parts list work package contains only those parts coded "F" in the third position of the SMR code, therefore, there may be a break in the item number sequence.

HOW TO LOCATE REPAIR PARTS

1. When NSNs or Part Numbers Are Not Known.

First. Using the table of contents, determine the assembly group to which the item belongs. This is necessary since figures are prepared for assembly groups and subassembly groups, and lists are divided into the same groups.

Second. Find the figure covering the functional group or the subfunctional group to which the item belongs.

Third. Identify the item on the figure and note the number(s).

Fourth. Look in the repair parts list work packages for the figure and item numbers. The NSNs and part numbers are on the same line as the associated item numbers.

2. When NSN Is Known.

First. If you have the NSN, look in the STOCK NUMBER column of the NSN index work package. The NSN is arranged in NIIN sequence. Note the figure and item number next to the NSN.

Second. Turn to the figure and locate the item number. Verify that the item is the one you are looking for.

3. When Part Number Is Known.

First. If you have the part number and not the NSN, look in the PART NUMBER column of the part number index work package. Identify the figure and item number.

Second. Look up the item on the figure in the applicable repair parts list work package.

ABBREVIATIONS

Abbreviation	Explanation

BOI Basis of Issue

MAC Maintenance Allocation Chart

NIIN National Item Identification Number (consists of the last 9-digits of the NSN) NSN National Stock Number

RPSTL Repair Parts and Special Tools List

SMR Source, Maintenance and Recoverability Codes

SRA Specialized Repair Activity

TMDE Test, Measurement and Diagnostic Equipment

UOC Usable on Code

END OF WORK PACKAGE

FIELD AND SUSTAINMENT MAINTENANCE

DIESEL ENGINE, FOUR-CYLINDER, FOUR CYCLE, FUEL INJECTED C-240PW-28

GROUP 01 COOLING SYSTEM: THERMOSTAT AND WATER PUMP

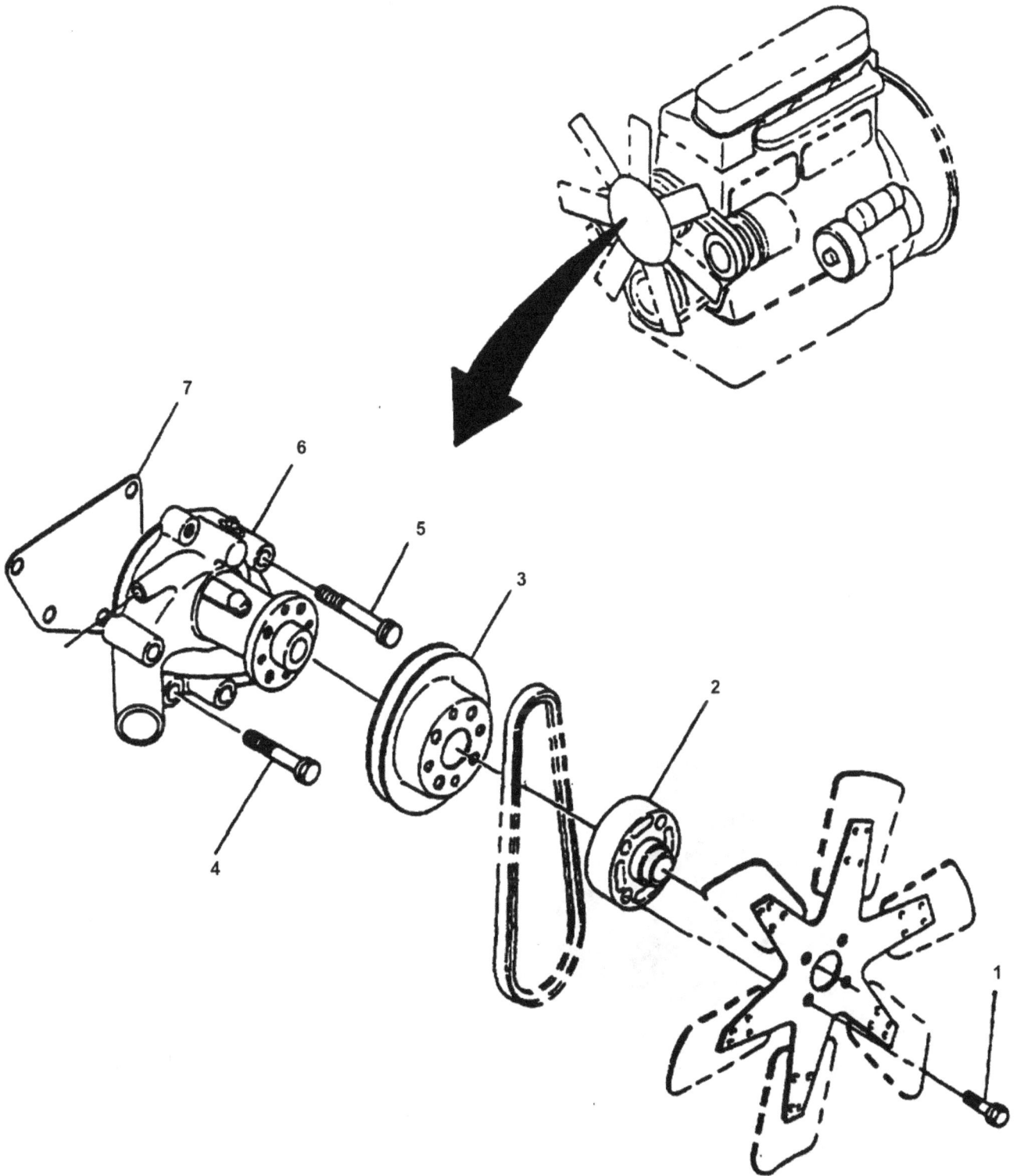

Figure 1. Thermostat and Water Pump (Sheet 1 of 2).

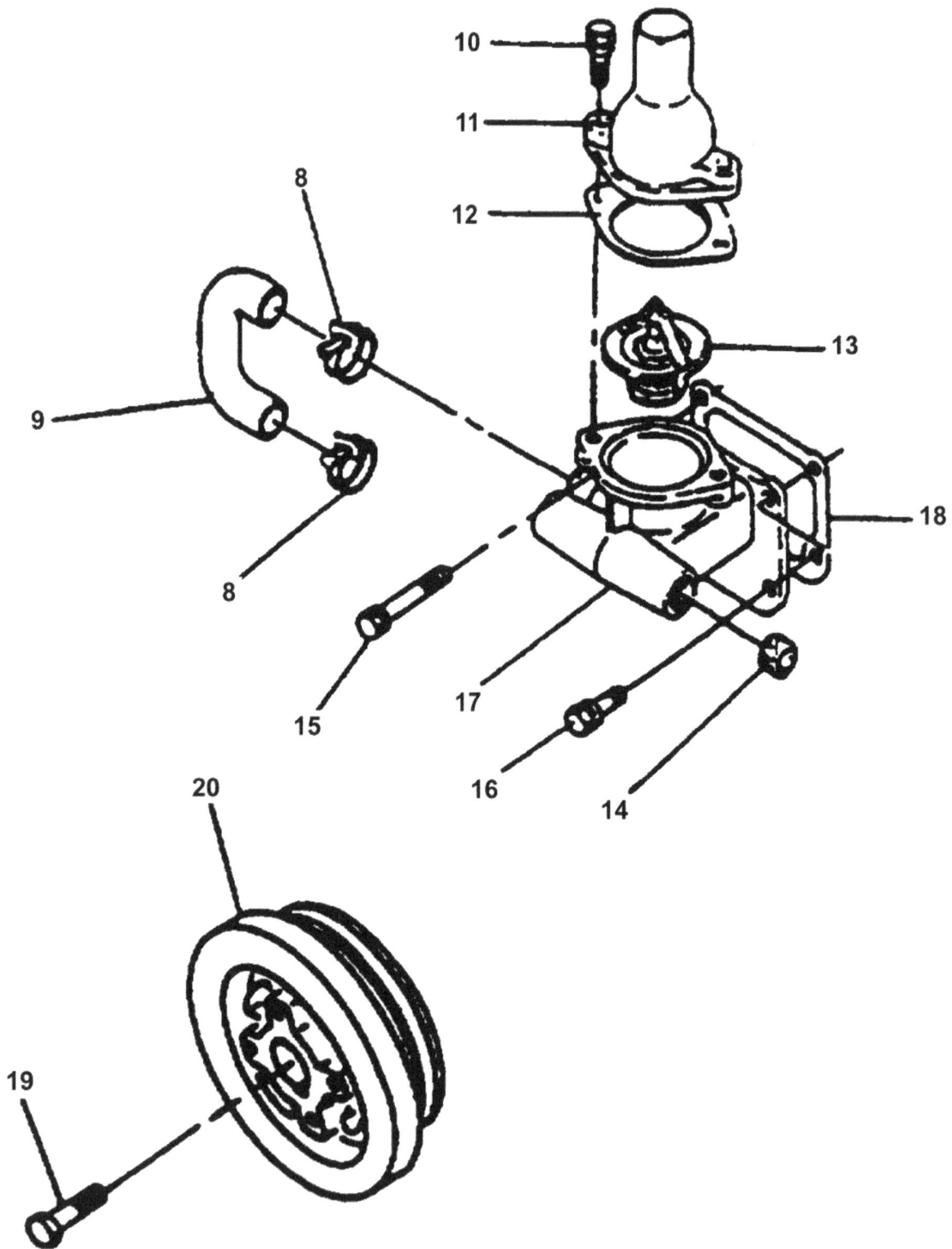

Figure 1. Thermostat and Water Pump (Sheet 2 of 2).

(1)	(2) SMR CODE		(3)	(4)	(5)	(6)	(7)
	a.	b.					
ITEM NO	ARMY	AIR FORCE	NSN	CAGEC	PART NUMBER	DESCRIPTION AND USABLE ON CODE (UOC)	QTY

GROUP 01 COOLING SYSTEM

FIG. 1 THERMOSTAT AND WATER PUMP

1 PAFZZ PAOZZ 5306-01-361-8359 S0226 0500408450 .BOLT,MACHINE 4

2 XDFZZ XB--- 4N530 9136420770 .SPACER,FAN 1

3 XDFZZ XB--- 4N530 9136411250 .PULLEY,FAN 1

4 XDFZZ XB--- 5306-01-376-0730 4N530 9137211322 .BOLT,ASSEMBLED WAS 4

5 PAFZZ PAOZZ 5305-01-268-0138 04627 324362 .SCREW, ASSEMBLED WAS .... 2 6 PAFZZ PAODD 2930-01-363-3096 S0226 897379805-0 .PUMP,COOLING SYSTEM WITH GAS-

KET 1

7 PAFZZ PAOZZ 5330-01-226-6695 04627 1557026 .GASKET, PART OF KIT P/N

587810-9401 1

8 PAFZZ PAOZZ 4730-01-363-5161 30076 0130845 .CLAMP,HOSE 2

9 PAFZZ PAOZZ 4720-01-382-2845 0LUY9 9137211322 .HOSE, PREFORMED 1

10 PAFZZ PAOZZ 5305-01-267-9279 04627 324373 .SCREW, ASSEMBLED WAS 2

11 PAFZZ PAOZZ 4710-01-363-2490 S0226 8971690151 .PIPE ASSEMBLY,METAL .... 1 12 PAFZZ PAOZZ 5330-01-413-3723 4N530 894362-2020 .GASKET, PART OF KIT P/N

587810-9401 1

13 PAFZZ PAOZZ 6620-01-220-7105 04627 324370 .THERMOSTAT 1

14 PAFZZ PAOZZ 4730-01-269-2313 S0226 5096050050 .PLUG,PIPE 1

15 PAFZZ PAOZZ 5306-01-224-9099 04627 324379 .BOLT 2

16 PAFZZ PAOZZ 5305-01-364-3142 4N530 0-50640-820-0 .SCREW,ASSEMBLED WAS 2

17 PBFZZ PBOZZ 2930-01-263-3162 30076 342741 .WATER OUTLET,ENGINE .... 1 18 PAFZZ PAOZZ 5330-01-226-4454 30076 324377 .GASKET, PART OF KIT P/N

587810-9401, KIT P/N 5878101862-2 1

19 XDFZZ XB--- 5306-01-361-8433 4F040 509000-132 .BOLT,MACHINE 1

20 XDFZZ XB--- 4N530 5123710901 .PULLEY, CRANK SHAFT 1

END OF FIGURE

FIELD AND SUSTAINMENT MAINTENANCE

DIESEL ENGINE, FOUR-CYLINDER, FOUR CYCLE, FUEL INJECTED C-240PW-28

GROUP 02 FUEL SYSTEM: FUEL FILTER ASSEMBLY

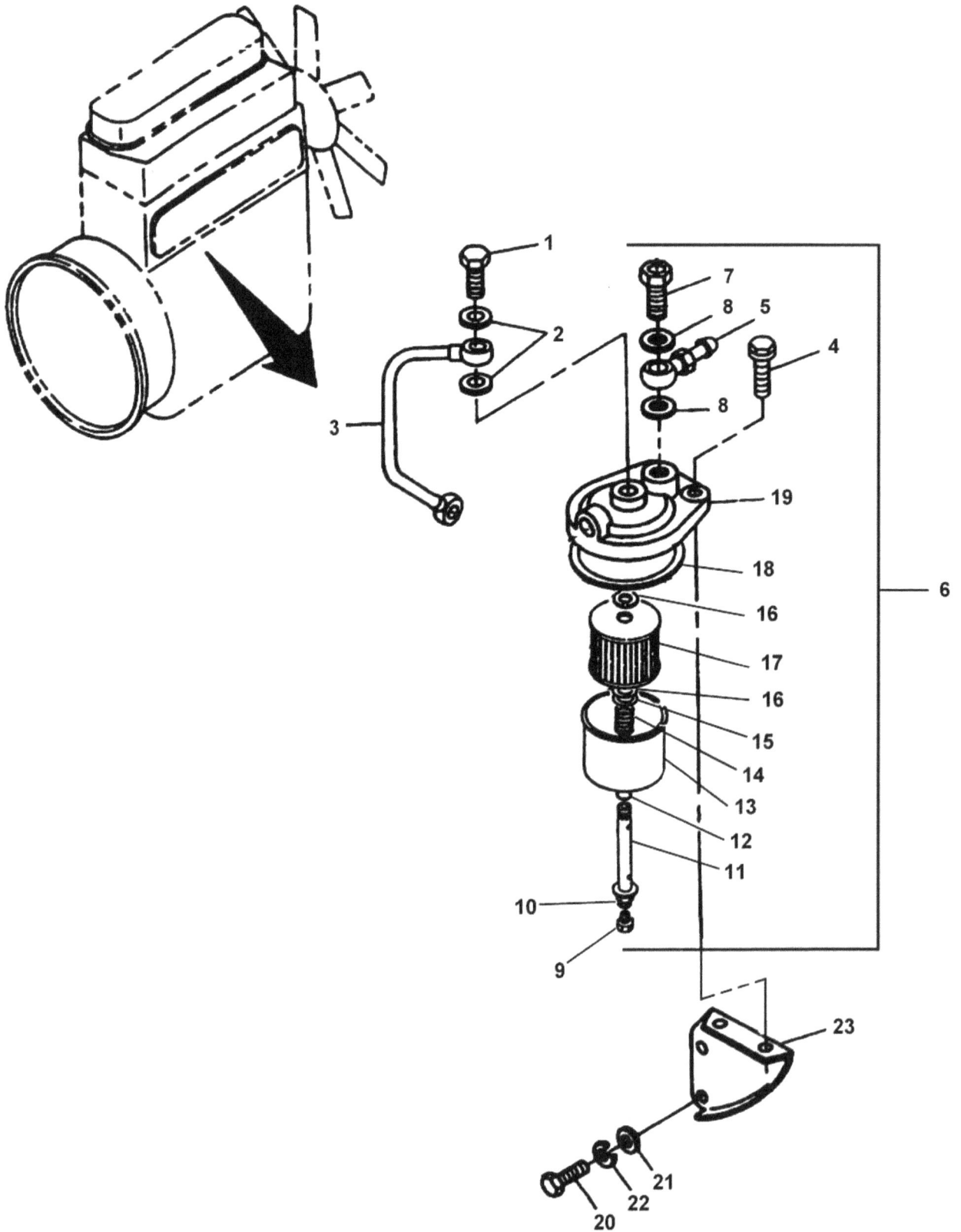

(1)	(2)		(3)	(4)	(5)	(6)	(7)
	SMR CODE						
	a.	b.					
ITEM NO	ARMY	AIR FORCE	NSN CAGEC PART NUMBER DESCRIPTION AND USABLE ON CODE (UOC)				QTY

GROUP 02 FUEL SYSTEM

FIG. 2 FUEL FILTER ASSEMBLY

1 PAFZZ PAOZZ 4730-01-280-4081 04627 324420 . BOLT,FLUID PASSAGE 1   2 PAFZZ PAOZZ 5330-01-224-9100 04627 324232 . SEAL, PART OF KIT P/N

587810-9401 2

3 PBFZZ PBOZZ 4710-01-560-9329 S0226 894402-6270 . PIPE ASSEMBLY,METAL 1

4 PAFZZ PAOZZ 5306-01-561-1338 S0226 060041-0320 . BOLT,ASSEMBLED WASH 2

5 PAFZZ PAOZZ 4730-01-221-7154 04627 324425 . CONNECTOR,MULTIPLE...................... 1

6 PBFOO PBOOO 2910-01-363-3087 S0226 9132018030 . FILTER,FLUID "NOT ILLUSTRATED" 1

7 XDFZZ XB-- 4N530 91332600562 . . VALVE,FLUID PRESSURE 1   8 PAFZZ PAOZZ 5330-01-224-9100 04627 324232 . . SEAL, PART OF KIT P/N

587810-9401 2

9 PAFZZ PAOZZ 5306-01-561-1246 4F040 913219-0070 . . BOLT,MACHINE 1   10 PAFZZ PAOZZ 5330-01-380-9155 S0226 9-0956-2103-0 . . O-RING, PART OF FUEL FILTER KIT,

P/N 587810-0090 1

11 XAFZZ XA--- 4N530 913231-0230 . . THRU-BOLT 1   12 PAFZZ PAOZZ 5330-01-380-9110 S0226 909920-5860 . . O-RING, PART OF FUEL FILTER KIT,

P/N 587810-0090 1

13 XAFZZ XA--- 4N530 913211-6040 . . BOWL, FILTER 1

14 PAFZZ PAOZZ 5330-01-560-9319 30076 909940-3500 . . SPRING,SPIRAL,TORSI 1

15 PAFZZ PAOZZ 5310-01-561-1396 30076 909851-3660 . . WASHER,LOCK 1   16 PAFZZ PAOZZ 5330-01-561-1363 S0226 909920-2430 . . GASKET, PART OF FUEL FILTER KIT

P/N 587810-0090 2

17 PAFZZ PAOZZ 4330-01-561-0833 S0226 UT6404 . . FILTER ELEMENT, FLUID, PART OF FUEL FILTER KIT P/N 587810-0090 1

18 PAFZZ PAOZZ 5330-01-382-5915 S0226 913213-0150 . . O-RING, PART OF FUEL FILTER KIT P/N 587810-0090 1

19 XAFZZ XA--- 4N530 913213-0150 . . BODY,FUEL FILTER 1

20 PAFZZ PAOZZ 5306-01-361-8434 4F040 9020608180 . BOLT,MACHINE 2

21 PAFZZ PAOZZ 5310-01-362-5946 S0226 9091645080 . WASHER,FLAT 2

22 PAFZZ PAOZZ 5310-01-265-2121 04627 373657 . WASHER,SPRING TENSI 2

23 PAFZZ PAOZZ 2990-01-263-3171 04627 324323 . BRACKET,ENGINE ACCE 1

END OF FIGURE

FIELD AND SUSTAINMENT MAINTENANCE

DIESEL ENGINE, FOUR-CYLINDER, FOUR CYCLE, FUEL INJECTED C-240PW-28

GROUP 02 FUEL SYSTEM: FUEL INJECTOR AND PIPING

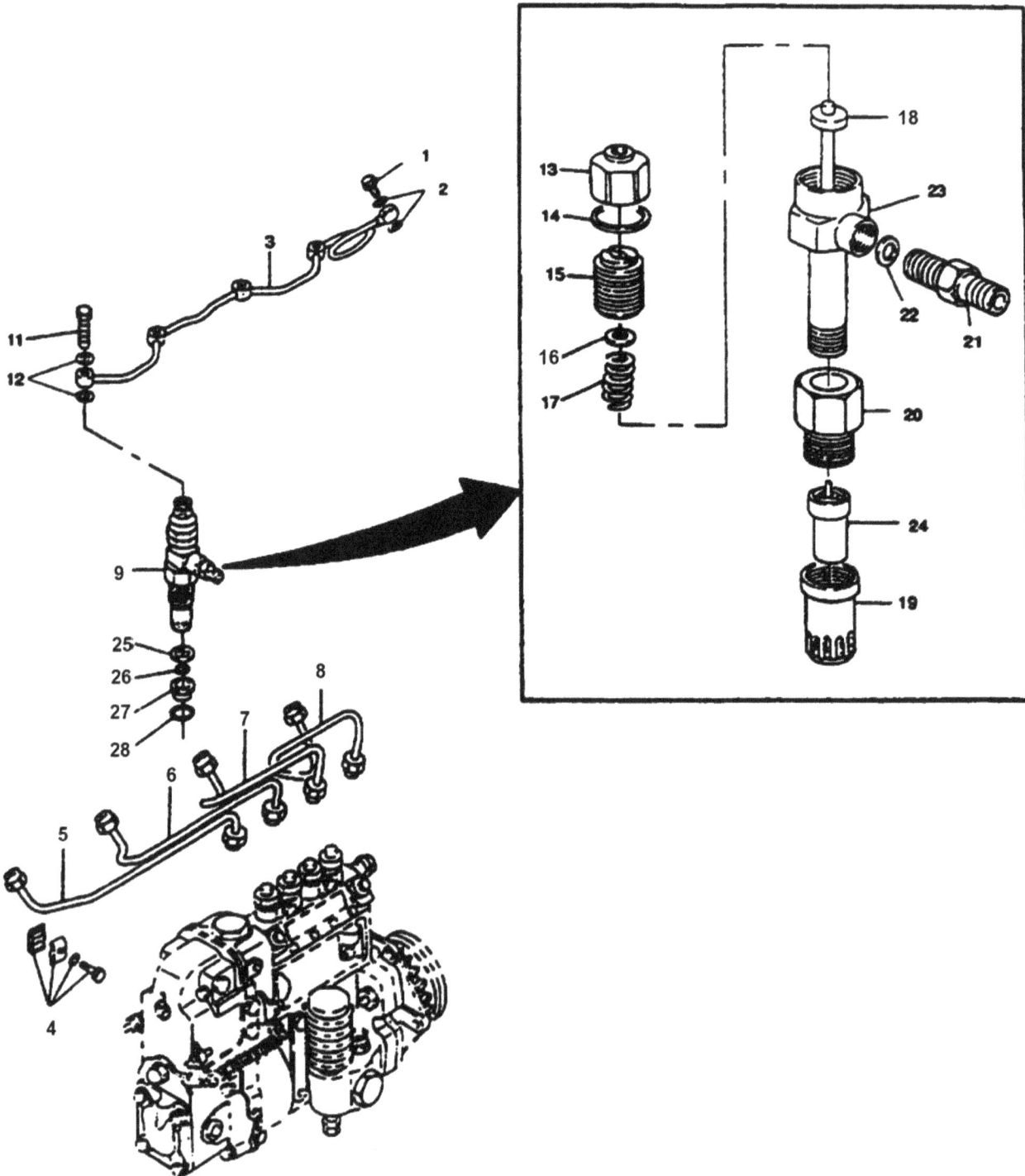

Figure 3. Fuel Injector and Piping.

(1)	(2)		(3)	(4)	(5)	(6)	(7)
	SMR CODE						
	a.	b.					
ITEM NO	ARMY	AIR FORCE	NSN	CAGEC	PART NUMBER	DESCRIPTION AND USABLE ON CODE (UOC)	QTY

GROUP 02 FUEL SYSTEM

FIG. 3 FUEL INJECTOR AND PIPING

1 PAFZZ PAOZZ 4730-01-222-7562 04627 324423 . BOLT, FLUID PASSAGE .... 1 2 PAFZZ PAOZZ 5330-01-224-9108 04627 324412 . SEAL, PART OF KIT PIN 587810-9401,

KIT P/N 5878101862-2 2

3 PAFZZ PAOZZ 4710-01-221-5770 04627 324422 . TUBE ASSEMBLY, METAL 1

4 PAFZZ PAOZZ 4030-01-272-7572 30076 324417 . CLAMP, WIRE ROPE, BOL .... 3 5 PAFZZ PAOZZ 4710-01-248-8113 S0226 515411-2191 . TUBE ASSEMBLY,METAL, FOR P/N

897081-4360 1

5 PAFZZ PAOZZ 4710-01-560-9321 S0226 894422-3561 . TUBE ASSEMBLY,METAL, FOR P/N
897190-3870 1

6 PAFZZ PAOZZ 4710-01-249-3576 S0226 515411-2180 . TUBE ASSEMBLY,METAL, FOR P/N
897081-4360 1

6 PAFZZ PAOZZ 4710-01-561-0841 S0226 884422-3551 . TUBE ASSEMBLY,METAL, FOR P/N
897190-3870 1

7 PAFZZ PAOZZ 4710-01-249-2119 S0226 515411-2171 . TUBE ASSEMBLY,METAL, FOR P/N
897081-4360 1

7 PAFZZ PAOZZ 4710-01-560-9892 S0226 894422-3541 . TUBE ASSEMBLY,METAL, FOR P/N
897190-3870 1

8 PAFZZ PAOZZ 4710-01-249-2118 S0226 515411-2160 . TUBE ASSEMBLY, METAL, FOR P/N
897081-4360 1

8 PAFZZ PAOZZ 4710-01-560-9882 S0226 884422-3531 . TUBE ASSEMBLY,METAL, FOR P/N
897190-3870 1

9 PAFFF PAOZZ 2910-01-361-0616 30076 1376449 . NOZZLE, FUEL INJECTION .... 4 10 XAFFF XA--- 4N530 8941145150 . . NOZZLE HOLDER "NOT ILLUS-

TRATED" 1

11 XDFZZ XB--- 4N530 1096751390 . . . BOLT, FLUID PASSAGE .... 1 12 PAFZZ PAOZZ 5330-01-224-9108 04627 324412 . . . WASHER,FLAT, PART OF KIT P/N

2587810-9401, KIT P/N
5878101862-2 2

13 XAFZZ XA--- 4N530 515334-0030 . . . NUT, CAP 1
14 PAFZZ PAOZZ 5330-01-250-1390 S0226 915339-0560 . . . GASKET 1
15 XAFZZ XA--- 4N530 9153396010 . . . SCREW, ADJUSTING 1
16 XAFZZ XA--- 4N530 9153390570 . . . WASHER 1
17 XAFZZ XA--- 4N530 5153430020 . . . SPRING, NOZZLE 1
18 XAFZZ XA--- 4N530 5153410020 . . . ROD, PUSH 1
19 XAFZZ XA--- 4N530 9153326030 . . . NUT, RETAINER 1
20 XAFZZ XA--- 4N530 9153596010 . . . SCREW 1
21 PAFZZ PAOZZ 4730-01-362-5575 4N530 9153510070 . . . NIPPLE, PIPE 1
22 PAFZZ PAOZZ 5330-01-250-1391 S0026 915352-0050 . . . GASKET 1
23 XAFZZ XA--- 4N530 8941145150-1 . . . NOZZLE, HOLDER 1
24 XAFZZ XA--- 2910-01-217-3645 04627 324407 . . NOZZLE,FUEL INJECTION .... 1 25 PAFZZ PAOZZ 5310-01-226-6661 04627 990615 . WASHER, PART OF KIT P/N

587810-9401, KIT P/N
58781017862-2 4

26 PAFZZ PAOZZ 5310-01-224-9107 30076 324411 . WASHER, PART OF KIT P/N
4587810-9401, KIT
P/N5878101862-2 4

27 PAFZZ PAOZZ 2805-01-221-7364 61888 Z9153150270 . DEFLECTOR, DIRT AND 4

(1)	(2)	(3)	(4)	·	(5)	(6)	(7)
	SMR CODE						
	a. b.						
ITEM NO	ARMY AIR FORCE			NSN CAGEC PART NUMBER DESCRIPTION AND USABLE ON CODE (UOC)			QTY

28 PAFZZ PAOZZ 5310-01-226-6660 30076 324410 . WASHER 4

END OF FIGURE

FIELD AND SUSTAINMENT MAINTENANCE

DIESEL ENGINE, FOUR-CYLINDER, FOUR CYCLE, FUEL INJECTED C-240PW-28

GROUP 02 FUEL SYSTEM: FUEL INJECTION PUMP ASSSEMBLY

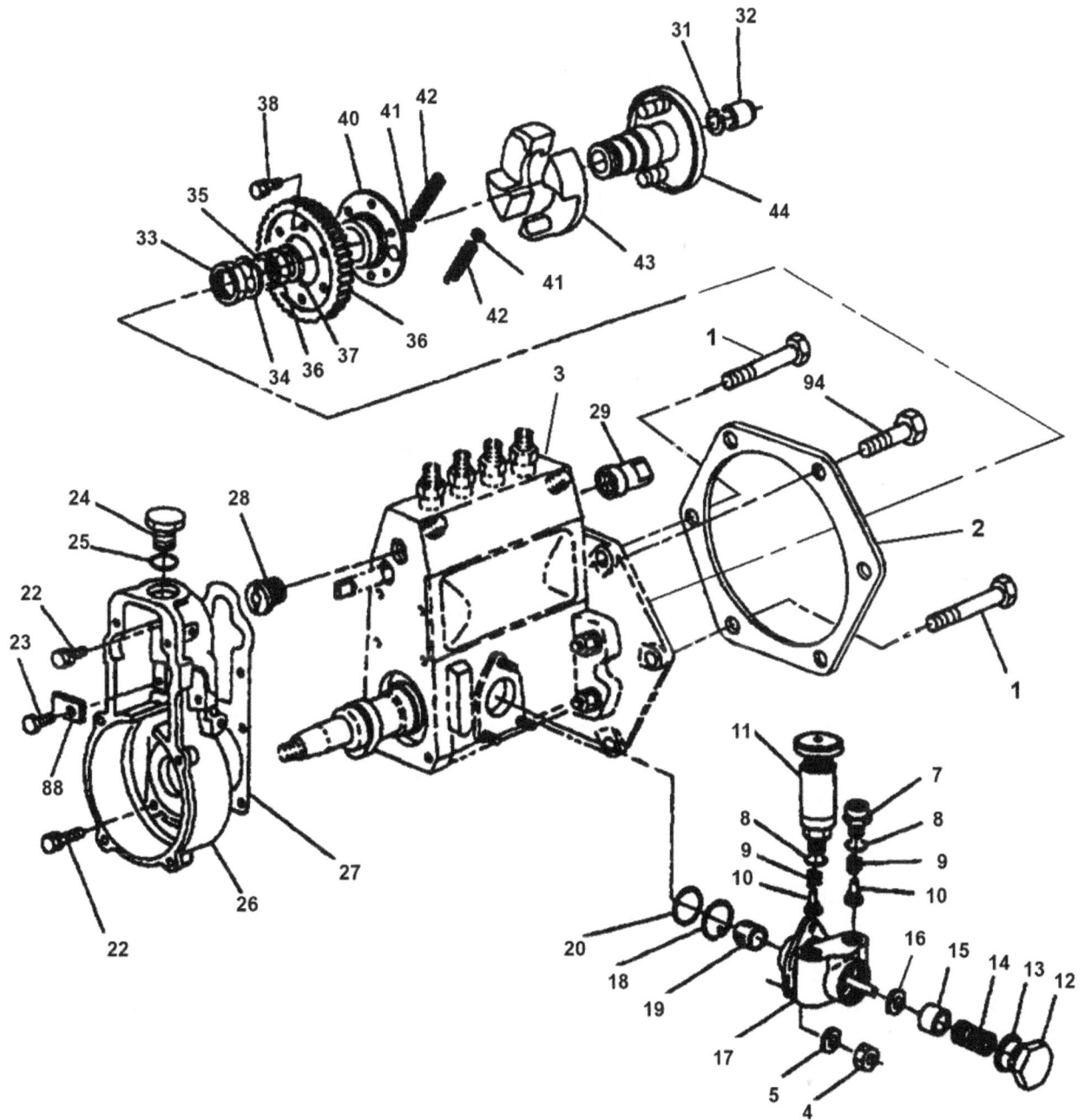

Figure 4. Fuel Injection Pump Asssembly (Sheet 1 of 2).

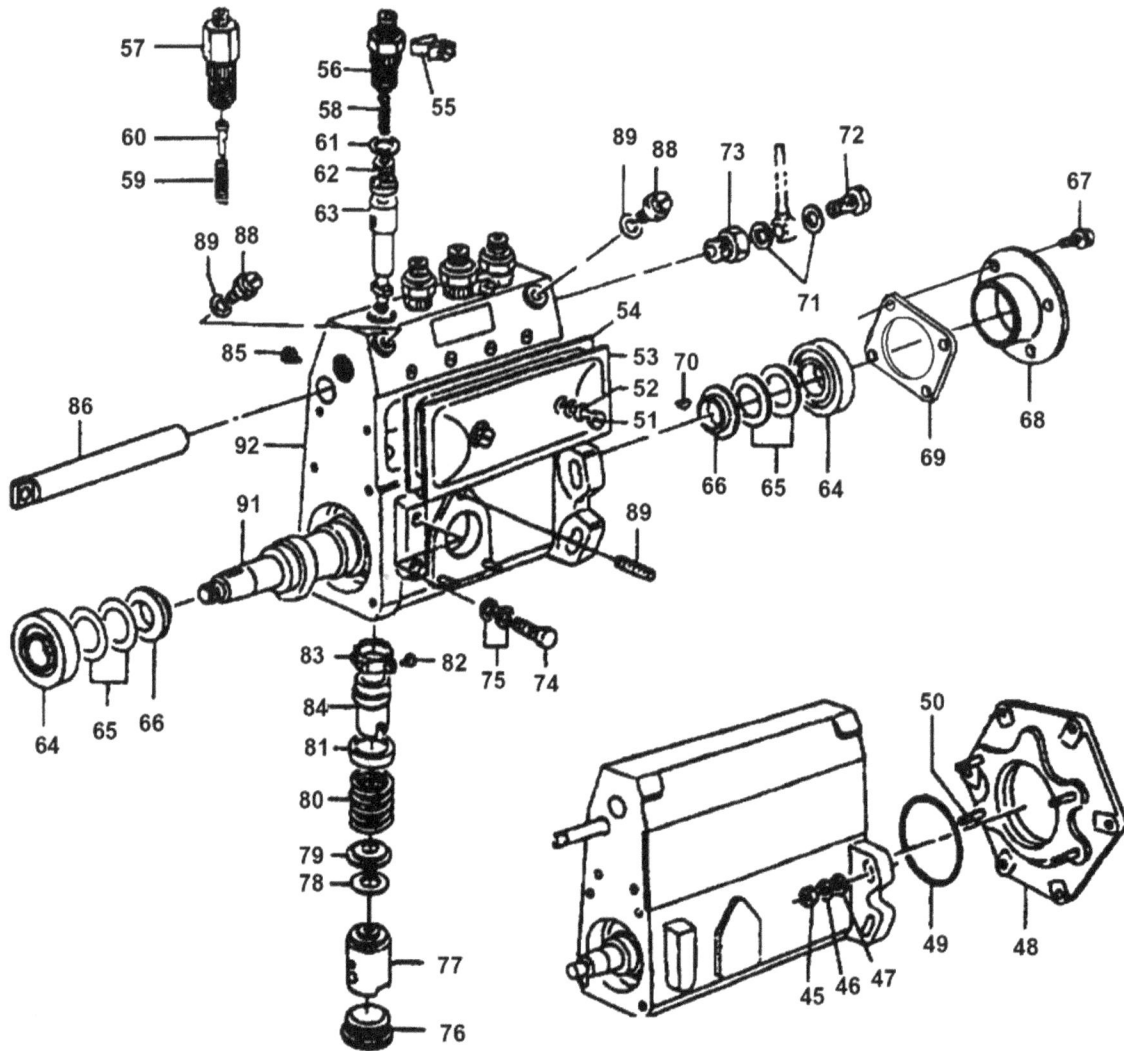

Figure 4. Fuel Injection Pump Asssembly (Sheet 2 of 2).

(1)	(2) SMR CODE a. b.	(3)	(4)	(5)	(6)	(7)
ITEM NO	ARMY AIR FORCE	NSN	CAGEC	PART NUMBER	DESCRIPTION AND USABLE ON CODE (UOC)	QTY

GROUP 02 FUEL SYSTEM

FIG. 4 FUEL INJECTION PUMP ASSSEMBLY

1 PAFZZ PAOZZ 5305-01-269-7534 S0226 8942059870 . SCREW, ASSEMBLED WASHER 2
2 PAFZZ PAOZZ 5330-01-265-2127 S0226 5113190570 . GASKET, PART OF KIT P/N 567610-9401 1
3 PAFHH PAOZZ 2910-01-384-5937 S0226 897081-4360 . PUMP,FUEL,METERING 1
3 PAFHH PAOZZ 2915-01-561-0259 30076 897190-3780 . PUMP,FUEL,METERING 1
4 PAFZZ XA--- 5310-01-362-5934 S0226 1157290060 .. NUT, PLAIN, HEXAGON 3
5 PAFZZ XA--- 5310-01-361-5723 S0226 109985179-0 .. WASHER,SPRING TENSI 3
6 PAFZF XA--- 2910-01-261-0112 04627 373969 . . PUMP,FEED "NOT ILLUSTRATED" 1
7 PAFZZ XA--- 4730-01-268-2448 04627 324570 . . . ADAPTER,CHECK VALVE 1
8 PAFZZ PAOZZ 5330-01-268-0159 04627 324571 . . . PACKING, PREFORMED 2
9 PBFZZ XA--- 5360-01-267-2931 04627 324568 . . . SPRING, HELICAL, COMP 2
10 PAFZZ XA--- 4820-01-264-5571 04627 324566 . . . VALVE, CHECK 2
11 PAFZZ XA--- 2910-01-251-2499 S0226 1157610061 . . . CONTROL HANDLE, PUMP 1
12 PAFZZ PAOZZ 5365-01-268-8993 04627 324563 . . . PLUG, MACHINE THREAD 1
13 PAFZZ PAOZZ 5330-01-267-9177 04627 324564 . . . GASKET 1
14 PAFZZ XA--- 5360-01-268-1019 04627 324562 . . . SPRING, HELICAL, COMP 1
15 PAFZZ XA--- 4320-01-264-5703 04627 324561 . . . PISTON, HYDRAULIC MO 1
16 PAFZZ XA--- 5330-01-561-0828 3PHX2 029620-4000 . . . PISTON, HYDRAULIC MO 1
17 XAFZZ XA--- 4N530 5157510120 . . . HOUSING, FEED PUMP 1
18 PAFZZ XA--- 5365-01-267-7498 30076 324572 . . . RING, RETAINING 1
19 PBFZZ XA--- 2910-01-264-8410 30076 324565 . . . TAPPET, FUEL PUMP 1
20 PAFZZ PAOZZ 5330-01-226-8686 4627 324449 . . . O-RING 1
21 XDFHH XA--- 4N530 5157201480 .. GOVERNOR ASSEMBLY "NOT ILLUSTRATED" 1
22 PAFZZ XA--- 5305-01-268-9066 30076 324504 . . . SCREW, CAP, HEXAGON H 6
23 PAFZZ XA--- 5305-01-270-1704 S0226 9019008180 . . . SCREW, ASSEMBLED WAS 1
24 XDFZZ XB--- 5340-01-556-2153 4N530 1157290380 . . . PLUG, HOUSING 1
25 PAFZZ PAOZZ 5331-01-416-2988 4N530 115729-0150 . . . O-RING 1
26 XDFZZ XB--- 4N530 1157210040 . . . HOUSING, GOVERNOR 1
27 PAFZZ PAOZZ 5330-01-226-8691 04627 324560 . . . GASKET 1
28 XDFZZ XB--- 4N530 9813150650 . . . ADAPTER, GOVERNOR 1
29 PAFZZ XA--- 5330-01-383-8706 S0226 9812350340 . . . CAP, SEAL, NONMETAL 1
30 PBHHH XA--- 3040-01-363-2360 16236 5157400720 . . . WEIGHT, COUNTERBALANCE "NOT ILLUSTRATED" 1
31 XDHZZ XA--- 4N530 1156190010 .. NUT, AUTO TIMER 1
32 XDHZZ XA--- 4N530 515619-0480 . . . WASHER, SPRING 1
33 XDHZZ XA--- 4N530 9813453060 . . . NUT, COUPLING FIX 1
34 XDHZZ XA--- 4N530 9813450520 . . . PLATE. LOCK 1
35 XDHZZ XA--- 4N530 515749-0070 . . . WASHER 1
36 XDHZZ XA--- 4N530 9813453250 . . . SHIM V
36 XDHZZ XA--- 4N530 9813453270 . . . SHIM V
36 XDHZZ XA--- 4N530 9813453280 . . . SHIM V
36 XDHZZ XA--- 4N530 9813453290 . . . SHIM V
36 XDHZZ XA--- 4N530 9813453300 . . . SHIM V
36 XDHZZ XA--- 4N530 9813453310 . . . SHIM V

(1)	(2) SMR CODE		(3)	(4)	(5)	(6)	(7)
	a.	b.					
ITEM NO	ARMY	AIR FORCE	NSN	CAGEC	PART NUMBER	DESCRIPTION AND USABLE ON CODE (UOC)	QTY
36	XDHZZ	XA---		4N530	9813453320	. . . SHIM	V
36	XDHZZ	XA---		4N530	9813453400	. . . SHIM	V
37	XDHZZ	XA---		4N530	9813453380	. . . WASHER	1
38	XDHZZ	XA---		4N530	5157490080	. . . SCREW, ASSEMBLED, W	6
39	PAHZZ	XA---	3020-01-264-5709	04627	324515	. . . GEAR, BEVEL	1
40	XDHZZ	XA---		4N530	9813433050	. . . FLANGE	1
41	XDHZZ	XA---		4N530	5157490090	. . . SHIM	V
41	XDHZZ	XA---		4N530	5157490100	. . . SHIM	V
41	XDHZZ	XA---		4N530	5157490110	. . . SHIM	V
41	XDHZZ	XA---		4N530	5157490680	. . . SHIM	V
41	XDHZZ	XA---		4N530	5157490690	. . . SHIM	V
41	XDHZZ	XA---		4N530	5157490700	. . . SHIM	V
41	XDHZZ	XA---		4N530	5157490710	. . . SHIM	V
41	XDHZZ	XA---		4N530	9813450430	. . . SHIM	V
41	XDHZZ	XA---		4N530	9813450440	. . . SHIM	V
41	XDHZZ	XA---		4N530	9813450450	. . . SHIM	V
42	XDHZZ	XA---		4N530	9813450720	. . . SPRING, TIMER	2
43	XDHZZ	XA---		4N530	9813435050	. . . FLYWEIGHT	2
44	XDHZZ	XA---		4N530	5157490280	. . . HOLDER, FLYWEIGHT	1
45	PAFZZ	XA---	5310-01-361-5702	S0226	5156390020	. NUT, PLAIN HEXAGON	4
46	PAFZZ	XA---	5310-01-361-5722	S0226	9091505100	. WASHER, LOCK	4
47	PAFZZ	XA---	5310-01-361-5718	S0226	9091606100	. WASHER, FLAT	4
48	PBFZZ	XA---	2990-01-264-3412	30076	324513	. BRACKET, ENGINE ACCE	1
49	PAFZZ	PAOZZ	5330-01-224-8063	30076	324514	. O-RING	1
50	PAFZZ	XA---	5307-01-381-2792	4F040	904411-030	. STUD, PLAIN	4
51	PAHZZ	XA---	5305-01-267-6310	04627	324446	. SCREW, CAP, HEXAGON H	2
52	PAHZZ	PAOZZ	5330-01-226-8700	04627	324447	. GASKET	2
53	XDHZZ	XA---		4N530	8941711090	. COVER, HOUSING SIDE	1
54	PAHZZ	XA---	5330-01-415-9615	4N530	894386-5960	. GASKET	1
55	XDHZZ	XA---		4N530	5156250010	. PLATE ASSEMBLY	2
56	PBHZZ	XA---	2530-01-264-3410	30076	359138	. BODY, VALVE, FOR P/N 897081-4360	4
57	XBHZZ	XA---		30076	131116-5920	. BODY,VALVE, FOR P/N 897190-3780	4
58	PAHZZ	XA---	5360-01-267-6712	04627	324452	. SPRING, HELICAL, COMP, FOR P/N 897081-4360	4
59	PAHZZ	XA---	5360-01-560-9887	30076	13112-0600	. SPRING, HELICAL, COMP, FOR P/N 897190-3780	4
60	PAHZZ	XA---	5340-01-561-0694	30076	13117-2500	. PLUG ASSEMBLY SEALING, FOR P/N 897190-3780	4
61	PAHZZ	PAOZZ	5330-01-361-8991	S0226	9812450140	. SEAL, PLAIN	4
62	PBHZZ	XA---	2910-01-265-0427	S0226	9812411030	. VALVE, INJECTION PUM	4
63	PBHZZ	XA---	3040-01-265-2608	30076	324450	. SHAFT, SHOULDERED	4
64	PAHZZ	XA---	3110-01-226-8680	04627	324490	. BEARING, BALL, ANNULA	1
64	PAHZZ	XA---	3110-01-226-8681	30076	324491	. BEARING, BALL, ANNULA	1
65	PAHZZ	XA---	5365-01-361-6567	1JY58	5156290180	. SHIM .16 MM	V
65	PAHZZ	XA---	5365-01-414-0745	1JY58	5156290900	. SHIM 1.40 MM	V
65	PAHZZ	XA---	5365-01-226-8764	30076	324494	. SHIM .12 MM	V
65	PAHZZ	XA---	5365-01-227-9883	04627	324495	. SHIM .14 MM	V
65	PAHZZ	XA---	5365-01-226-8766	04627	324497	. SHIM .16 MM	V
65	PAHZZ	XA---	5365-01-227-0633	04627	324498	. SPACER, SLEEVE SO MM	V
65	PAHZZ	XA---	5365-01-227-0612	30076	324501	. SHIM .70 MM	V

(1)	(2) SMR CODE		(3)	(4)	(5)	(6)	(7)
	a.	b.					
ITEM NO	ARMY	AIR FORCE	NSN CAGEC PART NUMBER DESCRIPTION AND USABLE ON CODE (UOC)				QTY

65	PAHZZ	XA---	5365-01-361-7899 3J868 5156190380 . SHIM 1.00 MM V
65	PAHZZ	XA---	5365-01-362-4406 3J868 5156290150 . SHIM .10 MM V
65	PAHZZ	XA--- 5365-01-467-5653 4N530 981225-0330 . SHIM .30 MM V
66	PAHZZ	XA---	5365-01-224-9109 04627 324492 . RING 2
67	PAHZZ	XA---	5305-01-268-9066 04627 324504 . SCREW, CAP, HEXAGON H 4
68	XDHZZ	XA--- 4N530 5156120120 . COVER, HOUSING 1
69	PAHZZ	XA--- 5330-01-560-9889 3PHX2 131321-1300 . GASKET 1
70	PAHZZ	XA--- 5315-01-467-3123 4N530 8-94221-411-0 . KEY, WOODRUFF 1
71	PBHZZ	XA---	2815-01-364-1519 30076 5156210170 . CAMSHAFT, ENGINE 1
72	PAFZZ	XA---	4730-01-264-3345 04627 373668 . BOLT, FLUID PASSAGE 1
73	PAFZZ	PAOZZ 5330-01-226-8690 S0226 157590010 . GASKET 2
74	PAFZZ	PAOZZ 5306-01-268-8982 S0226 9097060370 . BOLT, MACHINE 1
75	PAFZZ	XA--- 5330-01-361-7912 S0226 9095720100 . O-RING, PART OF KIT P/N 587810-9401 2
76	XDHZZ	XA--- 4N530 5156191360 . PLUG, TAPPET 4
77	PAHZZ	XA---	2910-01-264-5626 04627 324456 . TAPPET, FUEL PUMP 4
78	PAHZZ	XA---	5365-01-361-6568 1JY58 5156191080 . SHIM 1.35 MM V
78	PAHZZ	XA---	5365-01-361-6569 1JY58 5156191090 . SHIM 1.45 MM V
78	PAHZZ	XA---	5365-01-268-9115 04627 324457 . SHIM .30 MM V
78	PAHZZ	XA---	5365-01-269-1516 30076 324458 . SHIM .40 MM V
78	PAHZZ	XA---	5365-01-269-4450 04627 324459 . SPACER, RING .50 MM V
78	PAHZZ	XA---	5365-01-269-4453 30076 324460 . SPACER, RING .60 MM V
78	PAHZZ	XA---	5365-01-268-9151 04627 324461 . SPACER, RING .70 MM V
78	PAHZZ	XA---	5365-01-269-3783 04627 324462 . SPACER, RING .80 MM v
78	PAHZZ	XA---	5365-01-269-1049 04627 324463 . SPACER, RING .90 MM V
78	PAHZZ	XA---	5365-01-269-4452 04627 324464 . SPACER, RING 1.00 MM V
78	PAHZZ	XA---	5365-01-269-0043 04627 324465 . SPACER, RING 1.10 MM V
78	PAHZZ	XA---	5365-01-275-7181 04627 324466 . SPACER, RING 1.20 MM V
78	PAHZZ	XA---	5365-01-269-1050 04627 324467 . SPACER, RING 1.30 MM V
78	PAHZZ	XA---	5365-01-269-0042 04627 324468 . SPACER, RING 1.40 MM V
78	PAHZZ	XA---	5365-01-269-2714 04627 324469 . SPACER, RING .55 MM V
78	PAHZZ	XA---	5365-01-269-4449 04627 324470 . SPACER, RING .65 MM V
78	PAHZZ	XA---	5365-01-269-4448 04627 324471 . SPACER, RING .75 MM V
78	PAHZZ	XA---	5365-01-269-4451 04627 324472 . SPACER, RING .85 MM V
78	PAHZZ	XA---	5365-01-269-6280 30076 324473 . WASHER, FLAT .95 MM V
78	PAHZZ	XA---	5365-01-270-8920 04627 324474 . WASHER, FLAT 1.05 MM V
78	PAHZZ	XA---	5310-01-268-6976 04627 324475 . SPACER, RING 1.15 MM V
78	PAHZZ	XA---	5310-01-268-9750 04627 324476 . SPACER, RING 1.25 MM V
78	PAHZZ	XA---	5365-01-269-2713 04627 324477 . SPACER, RING .20 MM V
78	PAHZZ	XA---	5365-01-268-9114 04627 324479 . SHIM .35 MM V
78	PAHZZ	XA---	5365-01-269-0044 04627 324480 . SPACER, RING .45 MM V
78	PAHZZ	XA---	5365-01-361-6570 3J868 5156191100 . SHIM 1.55 MM V
78	PAHZZ	XA--- 5365-01-361-5724 S0226 5156190260 . SHIM .25 MM V
78	PAHZZ	XA--- 5365-01-361-5725 S0226 5156191060 . SHIM 1.50 MM V
78	PAHZZ	XA--- 5365-01-361-5726 S0226 5156191070 . SHIM 1.60 MM V
79	PAHZZ	XA---	5340-01-267-2963 04627 324463 . SEAT, HELICAL COMPRE 4
80	PAHZZ	XA--- 2910-01-363-2375 S0226 5156430090 . PLUNGER, FUEL INJECT 4
81	PAHZZ	PAOZZ 5340-01-268-2652 04627 324482 . SEAT, HELICAL COMPRE 4
82	PAHZZ	PAOZZ 5305-01-268-9758 30076 324486 . SCREW, CAP, HEXAGON H 4
83	PAHZZ	XA---	4320-01-264-8408 04627 324485 . PINION SLEEVE 4
84	PAHZZ	XA---	2805-01-264-3498 04627 324484 . CYLINDER SLEEVE 4
85	PAHZZ	PAOZZ 5305-01-267-9223 30076 324488 . SCREW, CAP, HEXAGON H 1

(1)	(2)		(3)	(4)	(5)	(6)	(7)
	SMR CODE						
	a.	b.					
ITEM NO	ARMY	AIR FORCE	NSN	CAGEC	PART NUMBER	DESCRIPTION AND USABLE ON CODE (UOC)	QTY

86 XDHZZ XA--- 4N530 5156370020 . RACK, CONTROL .... 1 87 XDHHH XDHHH 4N530 5156110460 . HOUSING ASSY, INJ "NOT ILLUS-

TRATED" 1

88 PAHZZ PAOZZ 5305-01-270-0073 04627 324443 . . SCREW, CAP, HEXAGON H 2

89 PAFZZ PAOZZ 5330-01-361-5287 S0226 8941445880 . . SEAL, PLAIN 2

90 PAFZZ PAOZZ 5307-01-266-9757 04627 324441 . . STUD, PLAIN 3

91 PAFZZ PAOZZ 4730-01-264-9460 04627 324442 . . ADAPTER, STRAIGHT, TU 1

92 XAHZZ XAHZZ 4N530 5156110460-6 . . HOUSING, FUEL PUMP 1

93 PAFZZ PAOZZ S0266 9813253970 . PLATE 1

94 PAFZZ PAOZZ 5360-01-361-8360 S0266 500408320 . BOLT, MACHINE 4

END OF FIGURE

FIELD AND SUSTAINMENT MAINTENANCE

DIESEL ENGINE, FOUR-CYLINDER, FOUR CYCLE, FUEL INJECTED C-240PW-28

GROUP 03 LUBRICATION SYSTEM: OIL FILTER AND PIPING

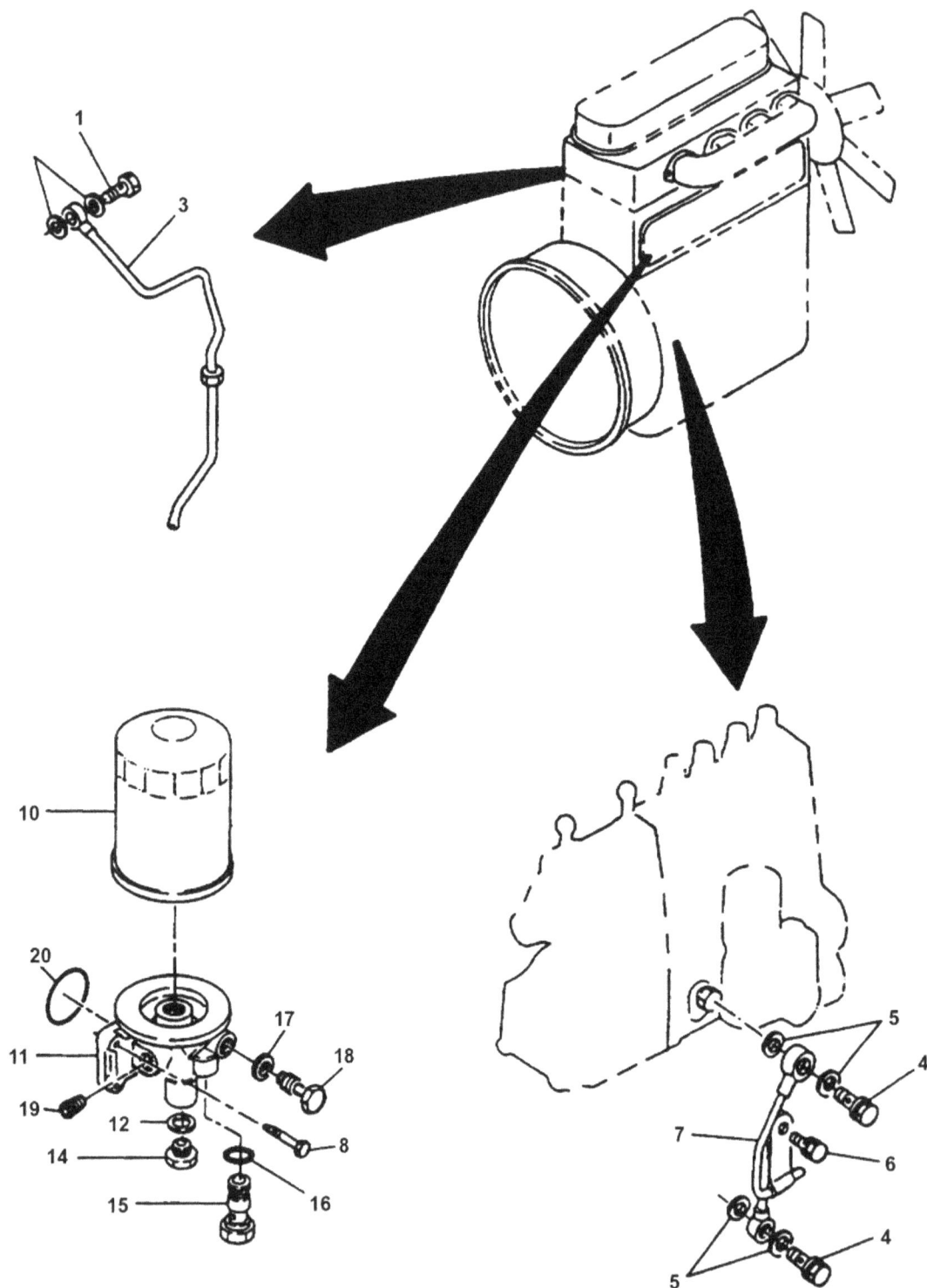

Figure 5. Oil Filter and Piping.

(1)	(2)	(3)	(4)	(5)	(6)	(7)
	SMR CODE					
	a. b.					
ITEM NO	ARMY AIR FORCE	NSN CAGEC PART NUMBER DESCRIPTION AND USABLE ON CODE (UOC)				QTY

GROUP 03 LUBRICATION
SYSTEM

FIG. 5 OIL FILTER AND PIPING

1 PAFZZ PAOZZ 4730-01-414-0348 4N530 1096750371 . BOLT, FLUID PASSAGE .... 1 2 PAFZZ PAOZZ 5330-01-414-3640 4N530 1096300820 . GASKET, PART OF KIT P/N

587810-9401 2

3 PBFZZ PBOZZ 4710-01-221-2214 04627 324351 . TUBE ASSEMBLY, METAL 1

4 PAFZZ PAOZZ 5306-01-226-9105 04627 324357 . BOLT .... 2 5 PAFZZ PAOZZ 5330-01-226-6692 04627 324356 . GASKET, PART OF KIT P/N

587810-9401 4

6 PAFZZ PAOZZ 5306-01-267-1778 S0226 9019006120 . BOLT, ASSEMBLED WASH 1

7 PBFZZ PBOZZ 4710-01-363-3041 S0226 5133111440 . PIPE, METALLIC 1

8 PAFZZ PAOZZ 5305-01-266-9343 S0226 901970845-0 . SCREW, ASSEMBLED WAS .... 4 9 PBFZZ PBOZZ 2910-01-360-6369 9M091 8941288541 . FILTER ELEMENT, FLUID "NOT

ILLUSTRATED" 1

10 PAFZZ PAOZZ 2940-01-493-4533 S0226 8-97024607-1 . . FILTER ELEMENT, FLUID 1

11 PBFZZ PBOZZ 2940-01-265-2668 30076 374896 . . HEAD, FLUID FILTER 1

12 PAFZZ PAOZZ 5330-01-414-3709 4N530 9099208410 . . SEAL, PLAIN, ENCASED 1

13 PAFZZ PAOZZ 4730-01-414-1259 4N530 913226-0130 . . ADAPTER, STRAIGHT, PI 1

14 XDFZZ XB--- 4N530 9098640150 . . PLUG TAPERED SCREW .... 1 15 PAFZZ PAOZZ 4820-01-232-5701 04627 324345 . . . VALVE, SAFETY RELIEF WITH

PACKING 1

16 PAFZZ PAOZZ 5330-01-383-6787 S0226 9099209750 . . . O-RING 1

17 PAFZZ PAOZZ 5330-01-413-3724 4N530 894167-8850 . O-RING 1

18 XDFZZ XB--- 4N530 8970887840 . PLUG 1

19 XDFZZ XB--- 4N530 8970387850 . PLUG .... 1 20 PAFZZ PAOZZ 5330-01-414-0691 3V577 897065-983-0 . O-RING, PART OF KIT P/N

587810-9401 1

END OF FIGURE

FIELD AND SUSTAINMENT MAINTENANCE

DIESEL ENGINE, FOUR-CYLINDER, FOUR CYCLE, FUEL INJECTED C-240PW-28

GROUP 03 LUBRICATION SYSTEM: OIL PAN ASSEMBLY

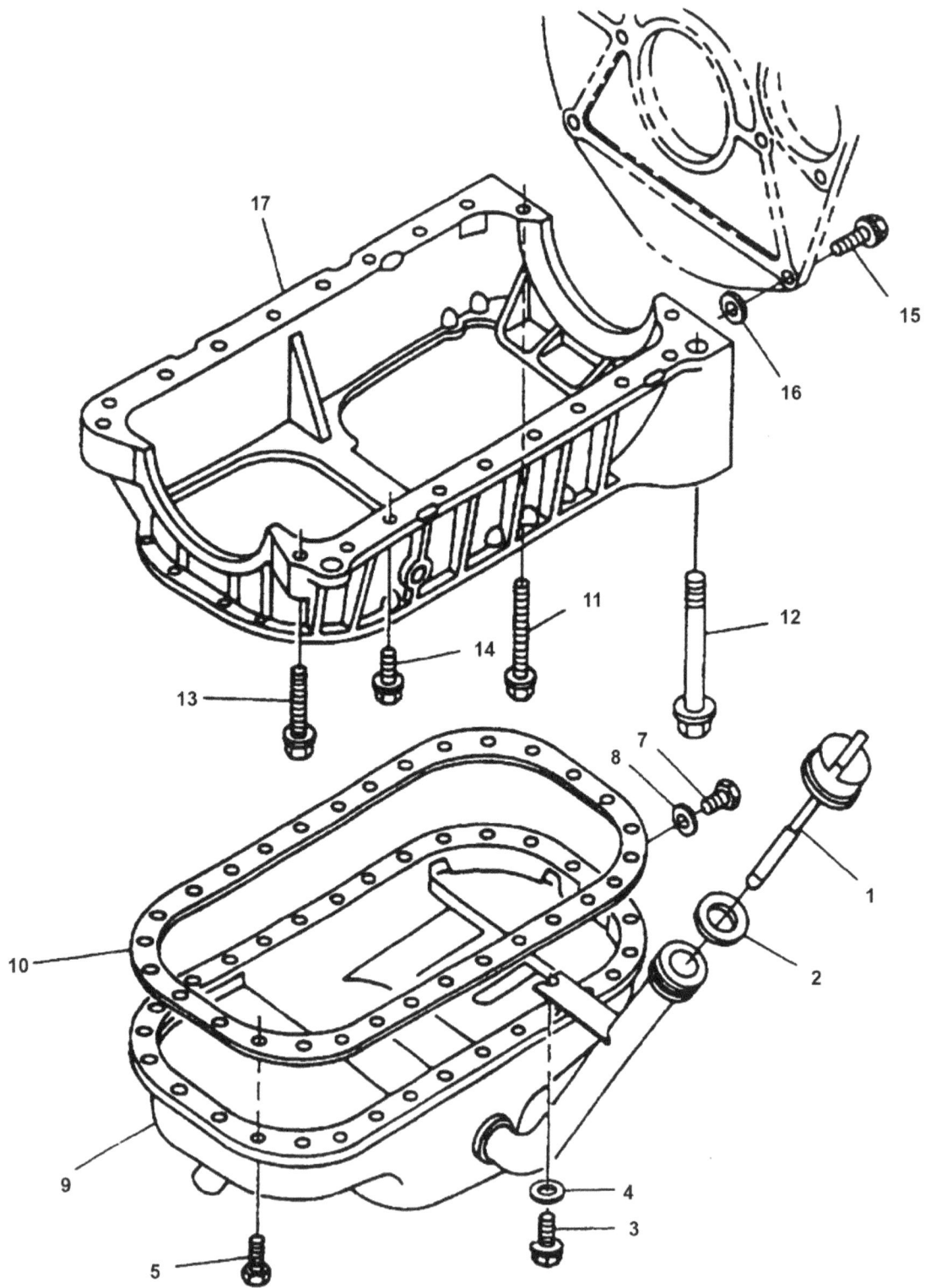

Figure 6. Oil Pan Assembly.

(1)	(2)		(3)	(4)	(5)	(6)	(7)
	SMR CODE						
	a.	b.					
ITEM NO	ARMY	AIR FORCE	NSN CAGEC PART NUMBER			DESCRIPTION AND USABLE ON CODE (UOC)	QTY

GROUP 03 LUBRICATION SYSTEM

FIG. 6 OIL PAN ASSEMBLY

1 XDFZZ XB--- 6680-01-472-9712 4N530 8970439250 . GAUGE ASSY, OIL LE 1

2 PAFZZ PAOZZ 5330-01-381-1934 4N530 909920-6090 . GASKET 1

3 PAFZZ PAOZZ 5306-01-361-8360 S0226 0500408320 . SCREW, CAP, HEXAGON H 1

4 PAFZZ PAOZZ 5310-01-360-5961 4F040 9091647080 . WASHER, FLAT 1

5 PAFZZ PAOZZ 5306-01-361-9044 4F040 050040-614 . BOLT, ASSEMBLED WASH 30

6 XAFZZ XA--- 4N530 5970367712-1 . PAN, OIL "NOT ILLUSTRATED" 1

7 PBFZZ PBOZZ 5365-01-414-3017 4F040 999202-3200 . . PLUG, MACHINE THREAD 1

8 PBFZZ PBOZZ 5330-01-361-6633 S0226 1096232350 . . GASKET 1

9 XDFFF XB--- 4N530 8970367712 . . PAN ASSEMBLY, OIL .... 1 10 PBFZZ PBOZZ 5330-01-224-8068 04627 324245 . GASKET, PART OF KIT P/N

587810-9401 1

11 PAHZZ PAOZZ 5306-01-272-9033 S0226 5090060870 . BOLT, ASSEMBLED WASH 2

12 PAHZZ PAOZZ 5306-01-414-3019 4N530 9-09801-202-2 . BOLT, MACHINE 2

13 PAHZZ PAOZZ 5305-01-266-9343 S0226 901970845-0 . SCREW, ASSEMBLED WAS 2

14 PAHZZ PAOZZ 5306-01-268-6796 04627 324249 . BOLT, MACHINE 14

15 PAHZZ PAOZZ 5305-01-381-2822 4F040 050041-232 . SCREW, MACHINE 2

16 PAHZZ PAOZZ 5365-01-265-4921 04627 324233 . SHIM .30 MM V

16 PAHZZ PAOZZ 5365-01-266-4081 04627 324234 . SHIM .35 MM V

16 PAHZZ PAOZZ 5365-01-265-4922 30076 324235 . SHIM .40 MM V

16 PAHZZ PAOZZ 5310-01-265-2109 04627 324236 . WASHER, FLAT .45 MM V

16 PAHZZ PAOZZ 5310-01-265-2110 04627 324237 . WASHER, FLAT .50 MM V

16 PAHZZ PAOZZ 5310-01-265-2111 04627 324238 . WASHER, FLAT .55 MM V

16 PAHZZ PAOZZ 5310-01-265-2112 04627 324239 . WASHER, FLAT .60 MM V

16 PAHZZ PAOZZ 5310-01-265-2113 04627 324240 . WASHER, FLAT .65 MM V

16 PAHZZ PAOZZ 5365-01-265-2114 04627 324241 . WASHER, FLAT .70 MM V

16 PAHZZ PAOZZ 5365-01-265-2115 04627 324242 . WASHER, FLAT .75 MM V

16 PAHZZ PAOZZ 5310-01-271-1003 04627 324243 . WASHER, FLAT .80 MM V

16 PAHZZ PAFZZ 5310-01-265-2116 04627 324244 . WASHER, FLAT .90 MM V

16 PAHZZ PAOZZ 5365-01-388-5092 S0226 9096511440 . SHIM .85 MM V

17 PBHZZ PBOZZ 2815-01-263-3150 04627 324230 . ENGINE BLOCK ASSEMBLY 1

END OF FIGURE

FIELD AND SUSTAINMENT MAINTENANCE

DIESEL ENGINE, FOUR-CYLINDER, FOUR CYCLE, FUEL INJECTED C-240PW-28

GROUP 03 LUBRICATION SYSTEM: OIL PUMP ASSEMBLY

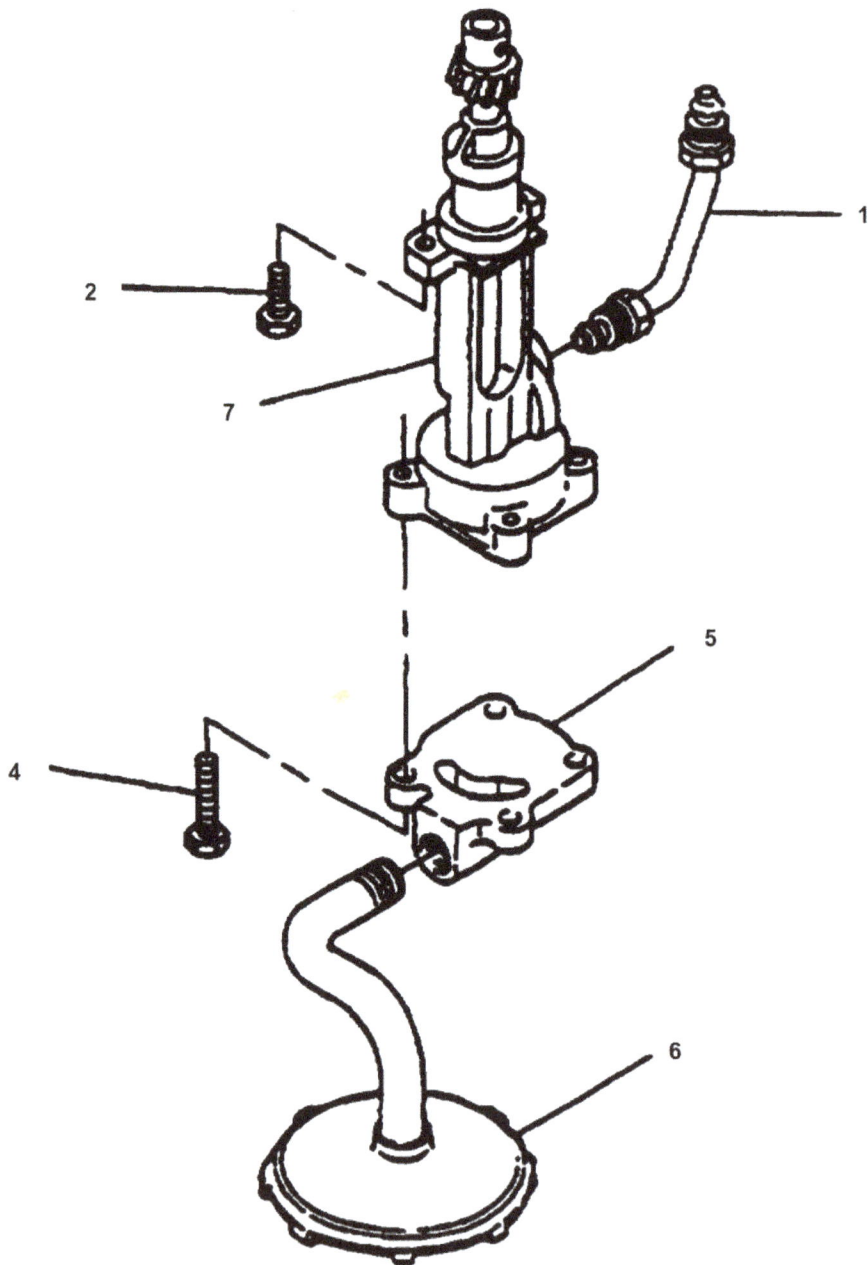

Figure 7. Oil Pump Assembly.

(1)	(2)		(3)	(4)	(5)	(6)	(7)
	SMR CODE						
	a.	b.					
ITEM NO	ARMY	AIR FORCE	NSN	CAGEC	PART NUMBER	DESCRIPTION AND USABLE ON CODE (UOC)	QTY

GROUP 03 LUBRICATION SYSTEM

FIG. 7 OIL PUMP ASSEMBLY

1 XDFZZ XB--- 4N530 8944548520 . PIPE ASSEMBLY, OIL 1
2 PAFZZ PAOZZ 5305-01-267-9222 30076 324263 . SCREW, CAP, HEXAGON HEAD .... 2 3 PAFZZ PAOZZ 2815-01-370-0208 S0226 9131008014 . OIL PUMP ASSEMBLY "NOT ILLUS-
TRATED" 1
4 PAFZZ PAOZZ 5305-01-267-8470 04627 373658 . . SCREW, CAP, HEXAGON HEAD 4
5 XAFZZ XA--- 4N530 9131130771 . . COVER, OIL PUMP 1
6 XAFZZ XA--- 4N530 9131401200 . . STRAINER ASSY, OIL 1
7 XAFZZ XA--- 4N530 5131200040 . . ROTOR ASSEMBLY 1

END OF FIGURE

FIELD AND SUSTAINMENT MAINTENANCE

DIESEL ENGINE, FOUR-CYLINDER, FOUR CYCLE, FUEL INJECTED C-240PW-28

GROUP 04 INTAKE AND EXHAUST SYSTEM: PVC ASSEMBLY

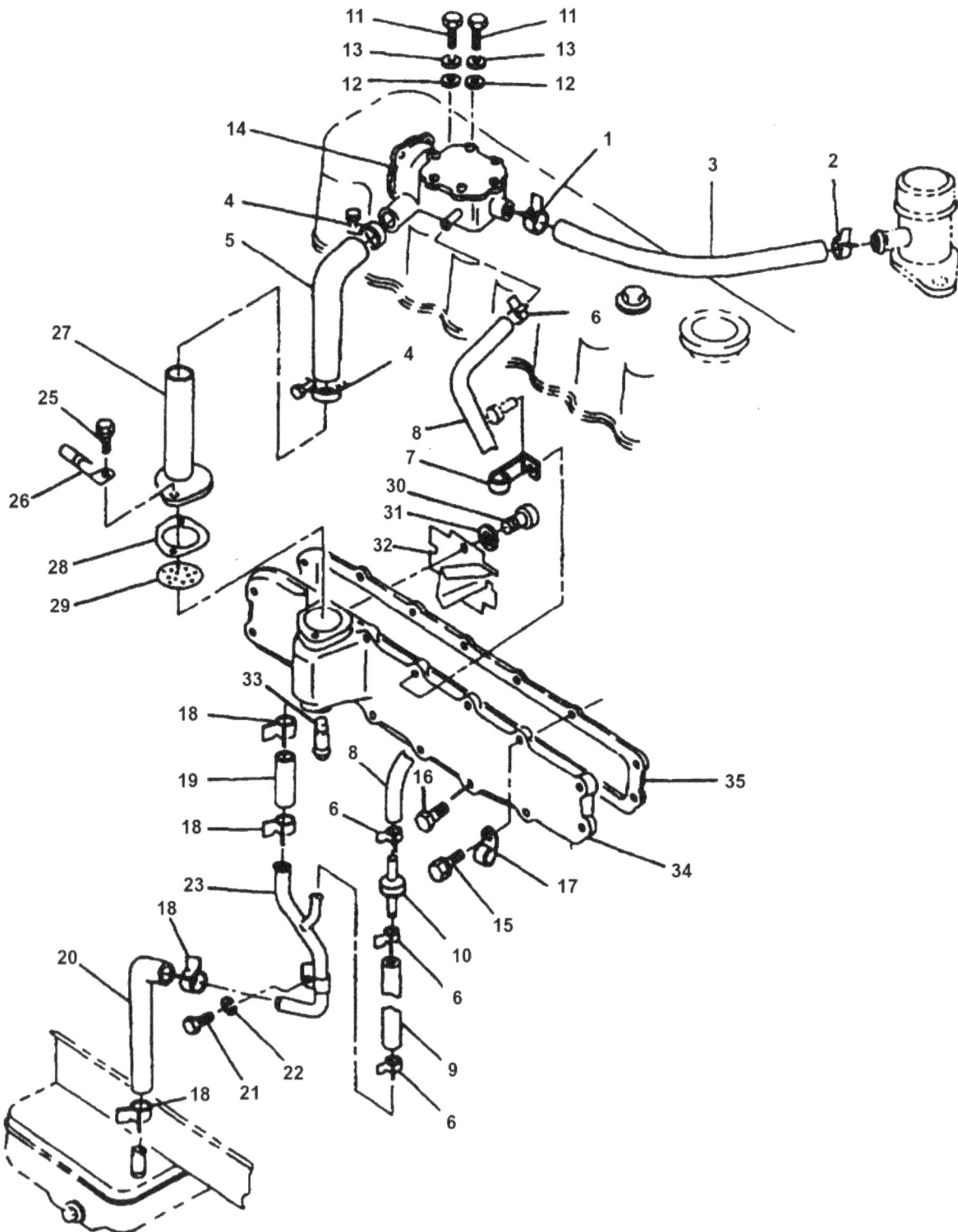

Figure 8. PVC Assembly.

(1)	(2)		(3)	(4)	(5)	(6)	(7)
	SMR CODE						
	a.	b.					
ITEM NO	ARMY	AIR FORCE	NSN	CAGEC	PART NUMBER	DESCRIPTION AND USABLE ON CODE (UOC)	QTY

GROUP 04 INTAKE AND
EXHAUST SYSTEM

FIG. 8 PVC ASSEMBLY

1 PAFZZ PAOZZ 4730-01-270-6432 04627 382507 . CLAMP, HOSE 1
2 PAFZZ PAOZZ 5340-01-361-8372 14153 9097151060 . STRAP, RETAINING 1
3 PAFZZ PAOZZ 4720-01-467-6675 4N530 8944225210 . HOSE, NONMETALLIC 1
4 PAFZZ PAOZZ 4730-01-263-7175 04627 382505 . CLAMP, HOSE 2
5 PAFZZ PAOZZ 4720-01-262-6271 04627 382504 . HOSE, PREFORMED 1
6 PAFZZ PAOZZ 4730-01-264-3880 04627 382502 . CLAMP, HOSE 4
7 PAFZZ PAOZZ 5340-01-272-4897 04627 382508 . CLAMP, LOOP .... 1 8 MOFZZ MOO-- 4N530 8941401600 . HOSE, NONMETALLIC MAKE FROM

P/N 1305789 (30076) 13 INCHES
REQUIRED 1

9 MOFZZ MOO-- 4N530 8944151560 . HOSE, NONMETALLIC MAKE FROM

P/N 1305789 (30076) 9 INCHES
. REQUIRED 1

10 PAFZZ PAOZZ 4820-66-128-5601 S0226 8942493171 . VALVE, CHECK 1
11 PAFZZ PAOZZ 5306-01-265-2062 S0226 9010456380 . BOLT, MACHINE 2
12 PAFZZ PAOZZ 5310-01-271-7712 04627 382511 . WASHER, FLAT 2
13 PAFZZ PAOZZ 5310-01-329-6975 04627 292680 . WASHER, LOCK 2
14 PBFZZ PBOZZ 4820-01-362-3399 S0226 8944682440 . VALVE ASSEMBLY 1
15 PAFZZ PAOZZ 5306-01-268-6796 04627 324249 . BOLT, MACHINE 2
16 PAFZZ PAOZZ 5305-01-270-1704 04627 324171 . SCREW, ASSEMBLED WASHER 12
17 PAFZZ PAOZZ 5340-01-227-3183 04627 324158 . CLIP 2
18 PAFZZ PAOZZ 4730-01-261-6818 30076 382498 . CLAMP, HOSE .... 4 19 MOFZZ MOO-- 4N530 22130800 . HOSE, NONMETALLIC MAKE FROM

P/N M6000-D0120 (81349), 4 INCHES
REQUIRED 1

20 PAFZZ PAOZZ 4720-01-261-4182 04627 382497 . HOSE, PREFORMED 1
21 PAFZZ PAOZZ 5306-01-361-8362 S0226 9-02041 016-0 . BOLT, MACHINE 1
22 PAFZZ PAOZZ 5310-01-361-5722 S0226 9091505100 . WASHER, LOCK 1
23 PBFZZ PBOZZ 4710-01-414-9304 4N530 8943172581 . TUBE, BENT, METALLIC .... 1 24 PAFZZ PAOZZ 2815-01-303-2505 04627 388291 . COVER, TAPPET "NOT ILLUS-

TRATED" 1

25 PAFZZ PAOZZ 5306-01-361-6607 4F040 9019606160 .. BOLT, ASSEMBLED WASHER 2
26 PAFZZ PAOZZ 5340-01-265-7276 04627 382488 .. CLIP, SPRING TENSION 1
27 PAFZZ PAOZZ 2590-01-263-3149 04627 382493 .. FILLER NECK 1
28 PAFZZ PAOZZ 5330-01-267-1231 04627 382490 .. GASKET 1
29 PBFZZ PBOZZ 2990-01-263-7230 04627 382487 .. FILTERING DISK, FLUID 1
30 PAFZZ PAOZZ 5305-01-267-9278 S0226 9934905100 .. SCREW, MACHINE 2
31 PAFZZ PAOZZ 5310-01-265-2118 04627 292679 .. WASHER, LOCK 1
32 PAFZZ PAOZZ 2805-01-263-6472 04627 382486 .. BAFFLE, AIRFLOW, ENGINE 1
33 XDFZZ XB--- 4N530 5096660100 .. NIPPLE, TAPPET COVER 1
34 XAFZZ XA--- 2815-01-263-3179 04627 382485 .. COVER, ENGINE POPPET .... 1 35 PAFZZ PAOZZ 5330-01-305-0239 04627 348442 . GASKET, PART OF KIT P/N

587810-9401 1

END OF FIGURE

FIELD AND SUSTAINMENT MAINTENANCE

DIESEL ENGINE, FOUR-CYLINDER, FOUR CYCLE, FUEL INJECTED C-240PW-28
GROUP 04 INTAKE AND EXHAUST SYSTEM

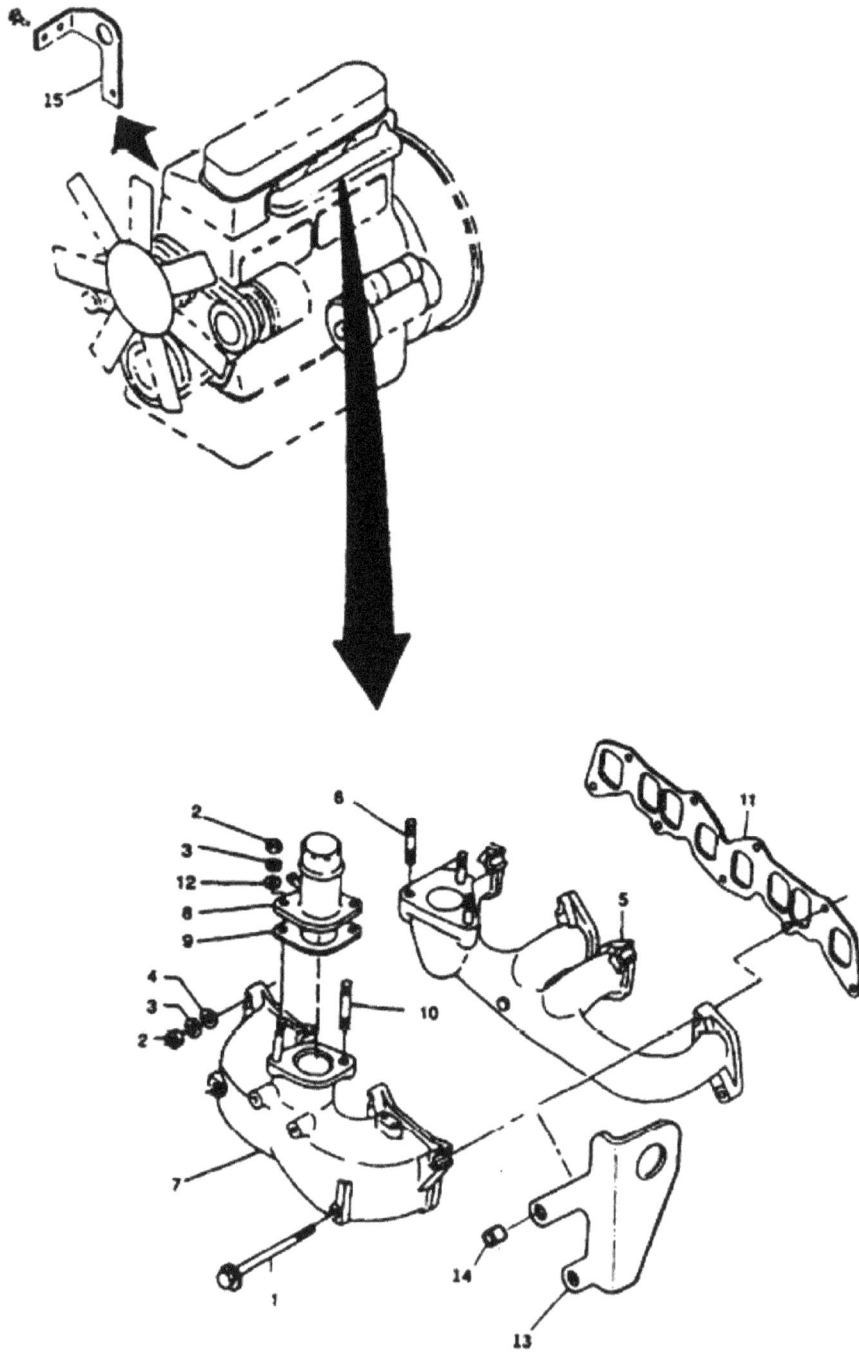

Figure 9. Intake and Exhaust System.

(1)	(2) SMR CODE a. b.	(3)	(4)	(5)	(6)	(7)
ITEM NO	ARMY AIR FORCE			NSN CAGEC PART NUMBER DESCRIPTION AND USABLE ON CODE (UOC)		QTY

GROUP 04 INTAKE AND EXHAUST SYSTEM

FIG. 9 INTAKE AND EXHAUST SYSTEM

1 PAFZZ PAOZZ 5306-01-272-9033 S0226 5090000870 . BOLT, ASSEMBLED WASHER 2

2 PAFZZ PAOZZ 5310-01-224-9104 04627 324314 . NUT 9

3 PAFZZ PAOZZ 5310-01-360-5966 4F040 109985-177 . WASHER, LOCK 9

4 PAFZZ PAOZZ 5310-01-360-7785 4F040 114319-034 . WASHER, FLAT 7

5 PBFZZ PBOZZ 2815-01-363-3098 04627 9141416143 . MANIFOLD, EXHAUST 1

6 PAFZZ PAOZZ 5307-01-381-2821 4F040 5093020020 . STUD, PLAIN 3

7 PBFZZ PBOZZ 2815-01-262-2760 04627 373661 . MANIFOLD, INTAKE 1

8 XDFZZ XB--- 5307-01-381-2821 4N530 5141111290 . PIPE, INLET .... 1 9 PAFZZ PAOZZ 5330-01-224-8071 04627 324397 . GASKET, PART OF KIT P/N

587810-9401 1

10 PAFZZ PAOZZ 5307-01-361-4612 S0226 904110-825 . STUD, PLAIN .... 2 11 PAFZZ PAOZZ 5330-01-224-8070 04627 1452558 . GASKET, PART OF KIT P/N

587810-9401, KIT P/N 5878101862-2 1

12 PAFZZ PAOZZ 5310-01-361-7892 1JY58 9091605080 . WASHER, FLAT .... 2 13 XDFZZ XB--- 4N530 9117771360 . BRACKET, ANGLE THIS BRACKET IS

NOT USED ON TACTICAL QUIET . GENERATORS 1

14 XDFZZ XB--- 5365-01-265-7125 04627 324437 . SPACER, SLEEVE 2

15 XDFZZ XB--- 2510-01-262-2650 04627 324435 . BRACKET, ENGINE MOUNT 1

END OF FIGURE

FIELD AND SUSTAINMENT MAINTENANCE

DIESEL ENGINE, FOUR-CYLINDER, FOUR CYCLE, FUEL INJECTED C-240PW-28

GROUP 05 ELECTRICAL SYSTEM: GLOW PLUGS

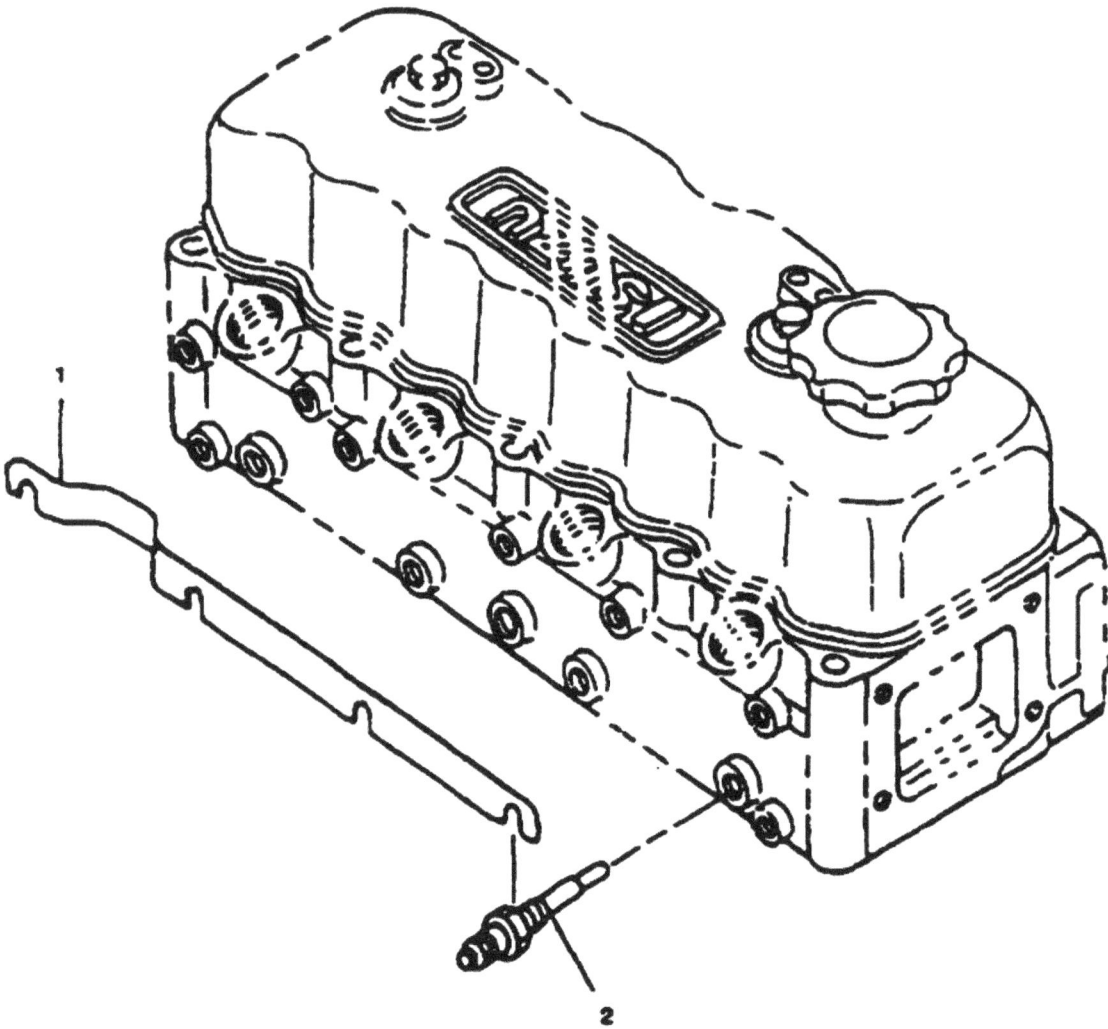

Figure 10. Glow Plugs.

(1)	(2)	(3)	(4)	(5)	(6)	(7)
	SMR CODE					
	a. b.					
ITEM NO	ARMY AIR FORCE		NSN CAGEC PART NUMBER DESCRIPTION AND USABLE ON CODE (UOC)			QTY

GROUP 05 ELECTRICAL SYSTEM

FIG. 10 GLOW PLUGS

1 PAFZZ PAOZZ 2990-01-262-2653 04627 324603 . BRACKET, ENGINE ACCE 1
2 PAFZZ PAOZZ 2920-01-360-3314 9M091 982513959 . GLOW PLUG 4

END OF FIGURE

FIELD AND SUSTAINMENT MAINTENANCE

DIESEL ENGINE, FOUR-CYLINDER, FOUR CYCLE, FUEL INJECTED C-240PW-28

GROUP 05 ELECTRICAL SYSTEM: STARTER

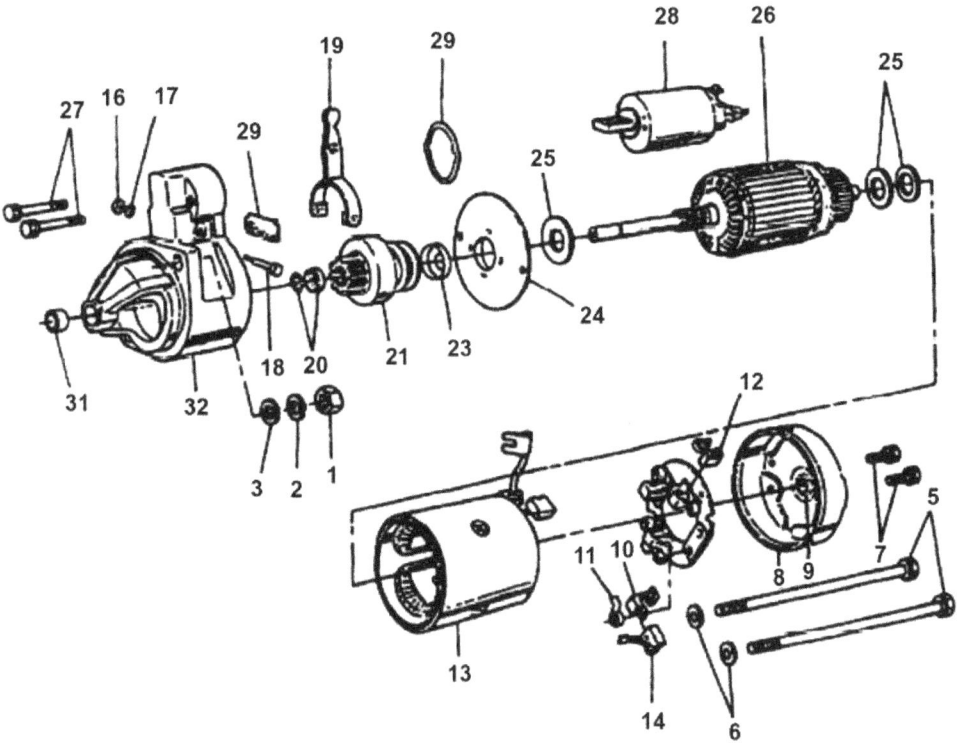

Figure 11. Starter.

(1)	(2)		(3)	(4)	(5)	(6)	(7)
	SMR CODE						
	a.	b.					
ITEM NO	ARMY	AIR FORCE	NSN	CAGEC	PART NUMBER	DESCRIPTION AND USABLE ON CODE (UOC)	QTY

GROUP 05 ELECTRICAL SYSTEM

FIG. 11 STARTER

1 PAFZZ PAOZZ 5310-01-414-4178 4N530 9991105120 . NUT, PLAIN, HEXAGON 2
2 PAFZZ PAOZZ 5310-01-414-3661 4N530 9091505120 . WASHER, LOCK 2
3 PAFZZ PAOZZ 5310-01-414-0741 4N530 9091606120 . WASHER, FLAT .... 2 4 PAFFF PAOOO 2920-01-385-5341 62786 S25-121A . MOTOR, ENGINE STARTI "NOT

ILLUSTRATED" 1
5 XDFZZ XB--- 62786 2250-3820 . . BOLT, THROUGH 2
6 PAFZZ PAOZZ 5310-01-376-2860 S0226 581129-0730 . . WASHER, FLAT 2
7 PAFZZ PAOZZ 5305-01-381-9841 62786 903910-4120 . . SCREW, ASSEMBLED, WA 2
8 PAFZZ PAOZZ 2920-01-363-4364 S0226 581115-013 . . HOUSING, ENGINE DRIV 1
9 PAFZZ PAOZZ 3120-01-250-2116 3W969 62-81501 . . . BEARING, SLEEVE 1
10 PAFZF PAFZF 5977-01-365-5584 S0226 581118-0020 . . HOLDER ASSEMBLY, ELE 1
11 PAFZZ PAOZZ 5360-01-560-9847 S0226 581117-0060 . . . SPRING, SPIRAL, TORSI 4
12 PAFZZ PAOZZ 5977-01-561-0860 3W969 68-8130 . . . BRUSH,ELECTRICAL CO 2
13 PAFZZ PAOZZ 2920-01-381-8693 30076 561112-0310 . . COIL ASSEMBLY, ELECT 1
14 PAFZZ PAOZZ 5977-01-561-0860 3W969 68-8130 . . BRUSH,ELECTRICAL CO .... 2 15 PAFZZ PAOZZ 3040-01-301-0405 30076 384091 . . LEVER, REMOTE CONTRO "NOT

ILLUSTRATED" 1
16 XAFZZ XA--- 62786 ST1 . . . NUT 1
17 XDFZZ XA--- 62786 ST2 . . . WASHER, LOCK 1
18 XDFZZ XA--- 62786 ST3 . . . STUD, PLAIN 1
19 XAFZZ XA--- 62786 ST4 . . . FORK, SHIFT 1
20 PAFZZ PAOZZ 5340-01-397-5931 4N530 581124-0010 . . . STOP, MECHANICAL 5
21 PAFZZ PAOZZ 2520-01-396-1397 4N530 581123-0170 . . . CLUTCH ASSEMBLY, FR .... 1 22 PAFZZ PAOZZ 2920-01-363-5181 9M091 5-81119-0130 . . END BELL, ELECTRICAL "NOT

ILLUSTRATED" 1
23 XAFZZ XA--- 3120-01-561-0957 3W969 62-81800 . . . BEARING,SLEEVE 1
24 XAFZZ XA--- 4N530 58119-0130 . . . BRACKET 1
25 PAFZZ PAOZZ 5330-01-360-9289 S0226 581129-0030 . . SEAL, PLAIN 3
26 PAFZZ PAOZZ 2920-01-363-5170 S0226 581121-0400 . . ARMATURE, MOTOR 1
27 PAFZZ PAOZZ 5305-01-395-1862 62786 050040-6450 . . SCREW, ASSEMBLED WAS 2
28 PAFZZ PAOZZ 2920-01-363-5178 S0226 5-81151058-1 . . RELAY-SOLENOID, ENGI 1
29 PAFZZ PAOZZ 5340-01-361-8415 14153 581129-0680 . . COVER, ACCESS .... 1 30 PAFZZ PAOZZ 2929-01-364-1622 30076 581113-0260 . . END BELL, ELECTRICAL "NOT

ILLUSTRATED" 1
31 PAFZZ PAOZZ 3120-01-560-9877 3W969 62-81502 . . . BEARING, SLEEVE 1
32 XAFZZ XA--- 4N530 581113-0260-1 . . . CASE 1

END OF FIGURE

FIELD AND SUSTAINMENT MAINTENANCE

DIESEL ENGINE, FOUR-CYLINDER, FOUR CYCLE, FUEL INJECTED C-240PW-28

GROUP 05 ELECTRICAL SYSTEM: ALTERNATOR

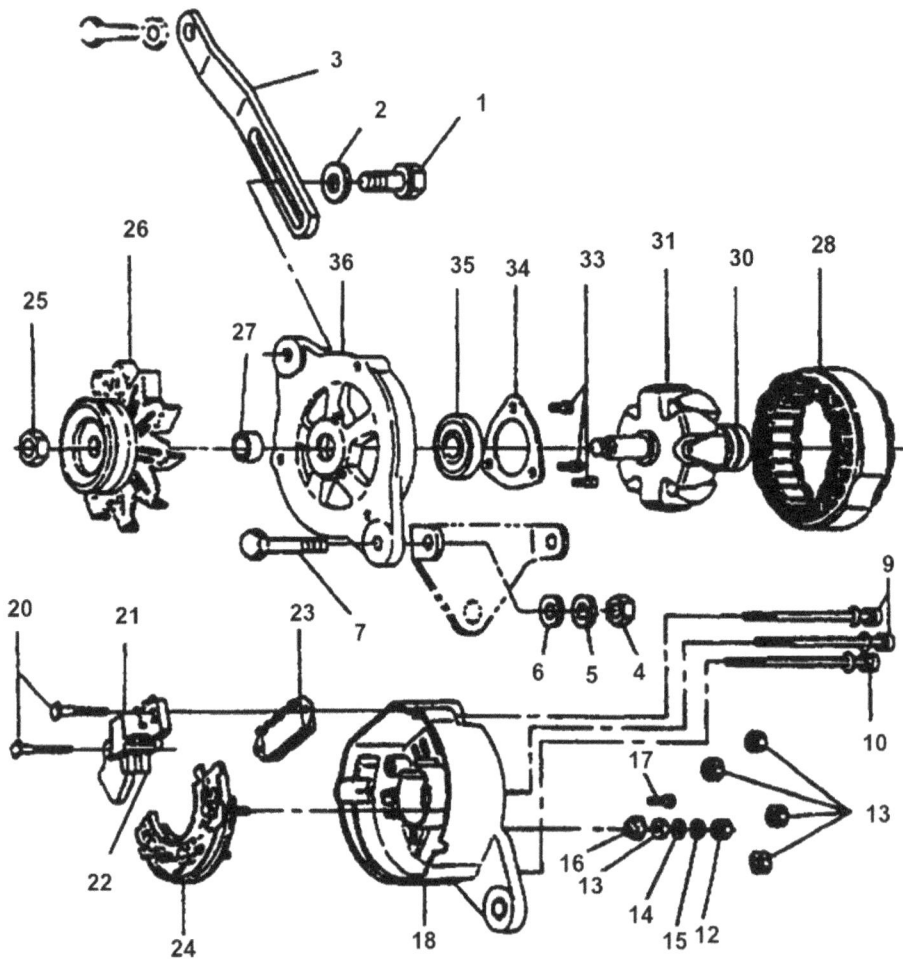

Figure 12. Alternator.

(1)	(2)		(3)	(4)	(5)	(6)	(7)
	SMR CODE						
	a.	b.					
ITEM NO	ARMY	AIR FORCE	NSN CAGEC PART NUMBER DESCRIPTION AND USABLE ON CODE (UOC)				QTY

GROUP 05 ELECTRICAL SYSTEM

FIG. 12 ALTERNATOR

1 PAFZZ PAOZ- 5306-01-224-9099 30076 324379 . BOLT 1
2 PAFZZ PAOZ- 5310-01-360-5961 4F040 9091647080 . WASHER, FLAT 1
3 XDFZZ XB--- S0226 5136740390 . BRACKET, ANGLE 1
4 PAFZZ PAOZ- 5310-01-381-0737 4F040 109440-040 . NUT, PLAIN, HEXAGON 2
5 PAFZZ PAOZ- 5310-01-360-5966 4F040 109985-177 . WASHER, LOCK 2
6 PAFZZ PAOZ- 5310-01-361-7892 1JY58 9091605080 . WASHER, FLAT 2
7 PAFZZ PAOZ- 5306-01-361-4586 4F040 901045-840 . BOLT, MACHINE 2
8 PAFFF PAOO- 6115-01-560-9146 62786 LR220-24 . GENERATOR, DIRECT CU 1
9 XDFZZ XB--- 62786 L125-6801 . . SCREW, ASSEMBLED, W 2
10 XDFZZ XB--- 62786 L125-7801 . . SCREW, ASSEMBLED, W 1
11 XAFFF XA--- 62786 L220-63001 . . COVER, REAR ASSY 1
12 PAFZZ PAOZ- 5310-01-168-8140 1FH08 DIN934M5 . . . NUT 1
13 PAFZZ PAOZ- 5310-01-383-8817 15526 DIN6926 M5 . . . NUT, SELF-LOCKING, EX 5
14 PAFZZ PAOZ- 5310-01-232-8491 80204 B1822BS050N . . . WASHER, FLAT 1
15 PAFZZ PAOZ- 5310-12-136-5792 D8286 DIN127-A5-FST-B3 ... WASHER, LOCK 1
16 XDFZZ XB--- 62786 NPT5 . . . INSULATOR 1
17 PAFZZ PAOZ- 5305-12-185-6163 D8286 DIN933M5X12A2 . . . CREW, CAP, HEXAGON H 1
18 XDFZZ XB--- 62786 L220-33011 . . . COVER, REAR 1
19 AFFFF AOOZ- 62786 L135-23152 . . . REGULATOR ASSEMBLY 1
20 PAFZZ PAOZ- 5306-01-382-4201 62786 L135-13134 . . . . BOLT, MACHINE 2
21 PAFZZ PAOZ- 5977-01-395-1865 4N530 581215-0460 . . . . HOLDER ASSEMBLY, ELE 1
22 PAFZZ PAOZ- 5977-01-560-9133 3W969 388102 . . . . BRUSH,ELECTRICAL CO 2
23 PAFZZ PAOZ- 2920-01-397-5289 33JP9 581251-0160 . . . . REGULATOR, ENGINE GE 1
24 PAFZZ PAOZ- 5961-01-376-4821 S0226 5-81262005-0 . . RECTIFIER, SEMICONDU 1
25 PAFZZ PAOZ- 5310-01-376-2850 62786 B-495469B . . NUT, PLAIN, EXTENDED 1
26 XDFZZ XB--- 62786 L215-55004 . . PULLEY FAN ASSEMBLY 1
27 XDFZZ XB--- 62786 NPT7 . . SLEEVE 1
28 PAFZZ PAOZ- 2920-01-362-5694 62786 L220-6200 . . STATOR ASSEMBLY, IGN 1
29 PAFFF PAOZ- 2920-01-397-1166 62786 L220-1100 . . ROTOR, GENERATOR 1
30 PAFZZ PAOZ- 3110-01-252-0446 S0226 894333-5210 . . . BEARING, BALL, ANNULA 1
31 XAFZZ XA--- 62786 NPT8 . . . ROTOR 1
32 XDFFF XB--- 62786 L120-14003 . . COVER, FRONT ASSY .... 1 33 PAFZZ PAOZ- 5305-01-560-9352 62786 L120-10001 . . . SCREW, ASSEMBLED, W4 MM X

0.7 MM X 12 MM 3

34 XDFZZ XB--- 62786 L135-14042 . . . RETAINER PLATE, BEA 1
35 PAFZZ PAOZ- 3110-00-144-8499 28873 BB5-21 . . . BEARING, BALL, ANNULA 1
36 XAFZZ XA--- 62786 ALT11 . . . COVER 1

END OF FIGURE

FIELD AND SUSTAINMENT MAINTENANCE

DIESEL ENGINE, FOUR-CYLINDER, FOUR CYCLE, FUEL INJECTED C-240PW-28

GROUP 06 CYLINDER HEAD AND CRANKCASE: FRONT GEAR COVER

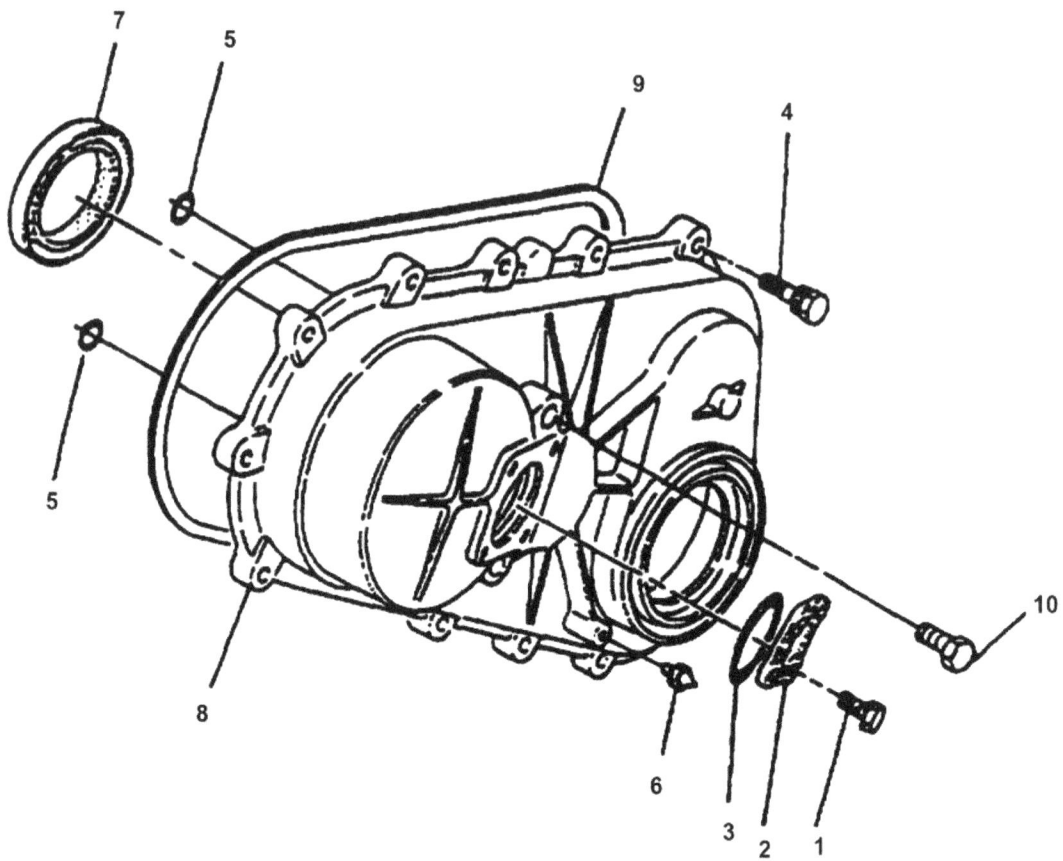

Figure 13. Front Gear Cover.

(1)	(2) SMR CODE		(3)	(4)	(5)	(6)	(7)
	a.	b.					
ITEM NO	ARMY	AIR FORCE	NSN	CAGEC	PART NUMBER	DESCRIPTION AND USABLE ON CODE (UOC)	QTY

GROUP 06 CYLINDER HEAD AND CRANKCASE

FIG. 13 FRONT GEAR COVER

(1)	(2a)	(2b)	(3)	(4)	(5)	(6)	(7)
1	PAFZZ	PAOZZ	5305-01-268-9066	30076	324504	. SCREW, CAP, HEXAGON H	4
2	XDFZZ	XB---		S0226	9113211480	. COVER, GEAR CASE	1
3	PAFZZ	PAOZZ	5330-01-361-8992	S0226	9099206010	. SEAL, PLAIN, PART OF KIT P/N 587810-9401	1
4	PAFZZ	PAOZZ	5306-01-361-8360	S0226	0500408320	. BOLT, MACHINE	13
5	PAFZZ	PAOZZ	5330-01-361-7914	S0226	9113120460	. O-RING, PART OF KIT P/N 587810-9401	2
6	PBFZZ	PBOZZ	5355-01-274-2456	04627	324203	. POINTER, DIAL	1
7	PAFZZ	PAOZZ	5330-01-226-6672	30076	355554	. SEAL, PLAIN ENCASED, PART OF KIT P/N 587810-9401	1
8	XDFZZ	XB---		S0226	5113110282	. CASE, GEAR, TIMING	1
9	PAFZZ	PAOZZ	5330-01-361-7913	S0226	9113120340	. O-RING, PART OF KIT P/N 587810-9401	1
10	PAFZZ	PAOZZ	5305-01-269-7534	S0226	8942059870	. SCREW, ASSEMBLED WAS	2

END OF FIGURE

FIELD AND SUSTAINMENT MAINTENANCE

DIESEL ENGINE, FOUR-CYLINDER, FOUR CYCLE, FUEL INJECTED C-240PW-28

GROUP 06 CYLINDER HEAD AND CRANKCASE: ROCKER COVER AND ARMS

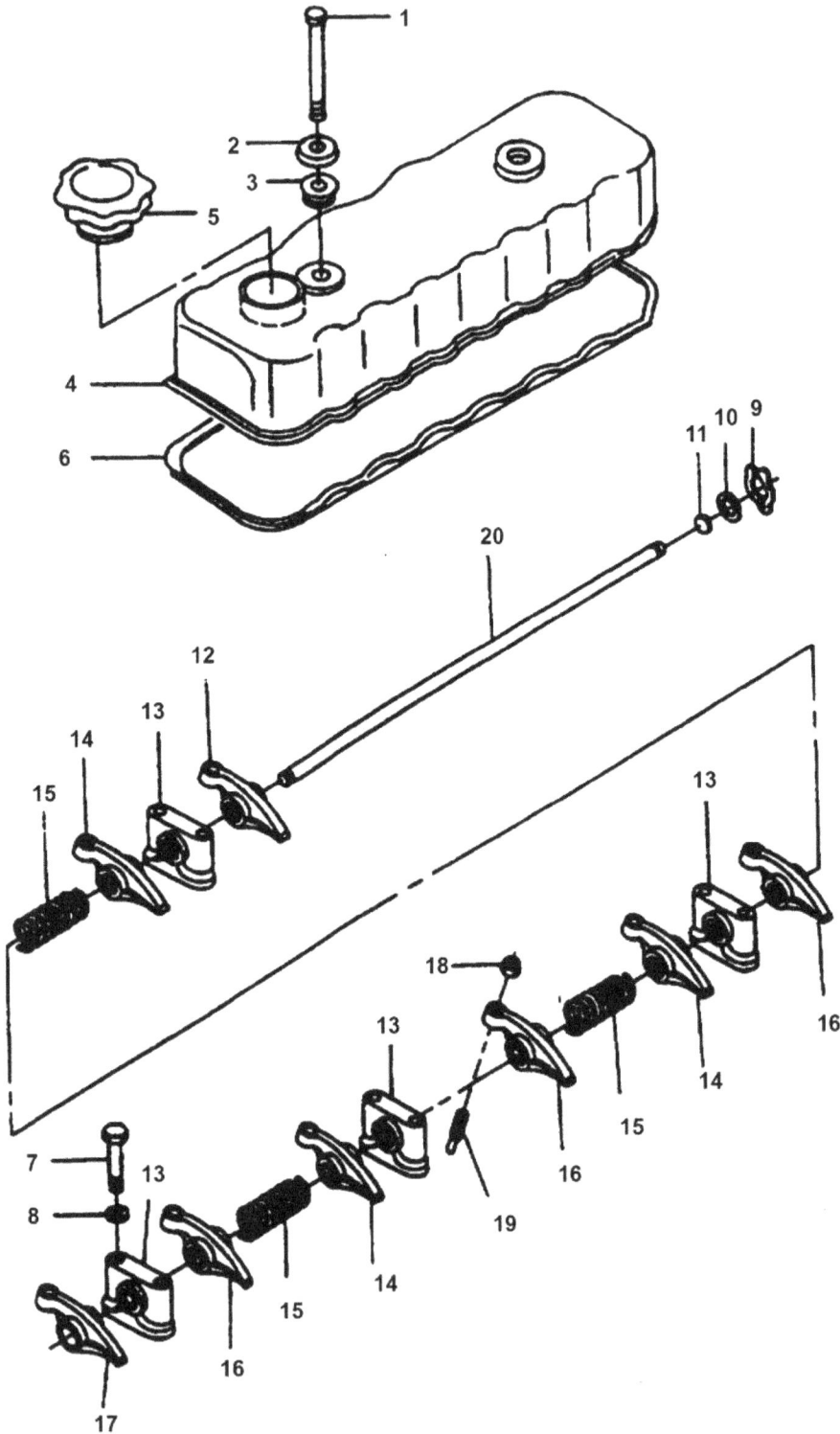

Figure 14. Rocker Cover and Arms.

(1)	(2)		(3)	(4)	(5)	(6)	(7)
	SMR CODE						
	a.	b.					
ITEM NO	ARMY	AIR FORCE	NSN	CAGEC	PART NUMBER	DESCRIPTION AND USABLE ON CODE (UOC)	QTY

GROUP 06 CYLINDER HEAD
AND CRANKCASE

FIG. 14 ROCKER COVER AND
ARMS

1 PAFZZ PAOZZ 5305-01-267-9286 S0226 5090000960 . SCREW, CAP, HEXAGON H 2
2 PBFZZ PBOZZ 5310-01-265-2120 04627 324135 . WASHER, RECESSED .... 2 3 PBFZZ PBOZZ 5330-01-226-4455 04627 324134 . GASKET, PART OF KIT P/N

                               587810-9401, KIT P/N
                               5878101862-2 2

4 PBFOO PBOZZ 2815-01-414-9430 4N530 8944054641 . COVER, ENGINE POPPET 1
5 PAFZZ PAOZZ 2590-01-260-0382 S0226 111751005-1 . CAP, FILLER OPENING .... 1 6 PAFZZ PAOZZ 5330-01-361-7915 S0226 8944475700 . O-RING, PART OF KIT P/N

                               587810-9401, KIT P/N
                               5878101862-2 1

7 PAFZZ PAOZZ 5306-01-102-2644 21969 MBX50DIN931-8.8 . BOLT, MACHINE 8
8 PAFZZ PAOZZ 5310-01-265-2108 04627 324259 . WASHER, FLAT 8
9 PAFZZ PAOZZ 5365-01-361-6807 S0226 9091800190 . RING, RETAINING 2
10 PAFZZ PAOZZ 5310-01-224-8007 04627 324321 . WASHER 2
11 PAFZZ PAOZZ 5340-01-265-2246 04627 324316 . PLUG, PROTECTIVE, DUST 2
12 PAFZZ PAOZZ 2815-01-222-5492 04627 324310 . ROCKER ARM, ENGINE P 1
13 PBFZZ PBOZZ 3040-01-222-4717 04627 324317 . PLATE, RETAINING, SHA 4
14 PAFZZ PAOZZ 2815-01-222-5491 04627 324311 . ROCKER ARM, ENGINE P 3
15 PAFZZ PAOZZ 5360-01-224-9102 30076 324318 . SPRING 3
16 PAFZZ PAOZZ 2815-01-221-8926 30076 324312 . ROCKER ARM, ENGINE P 3
17 PAFZZ PAOZZ 2815-01-221-8925 04627 324309 . ROCKER ARM, ENGINE P 1
18 PAFZZ PAOZZ 5310-01-360-5911 4F040 9091115080 . NUT, PLAIN, HEXAGON 8
19 XDFZZ XB--- 4N530 8941522830 . SCREW, ADJ. 8
20 PBFZZ PBOZZ 3040-01-222-3752 04627 348462 . SHAFT, ROCKER ARM 1

END OF FIGURE

FIELD AND SUSTAINMENT MAINTENANCE

DIESEL ENGINE, FOUR-CYLINDER, FOUR CYCLE, FUEL INJECTED C-240PW-28

GROUP 06 CYLINDER HEAD AND CRANKCASE: CYLINDER HEAD ASSEMBLY

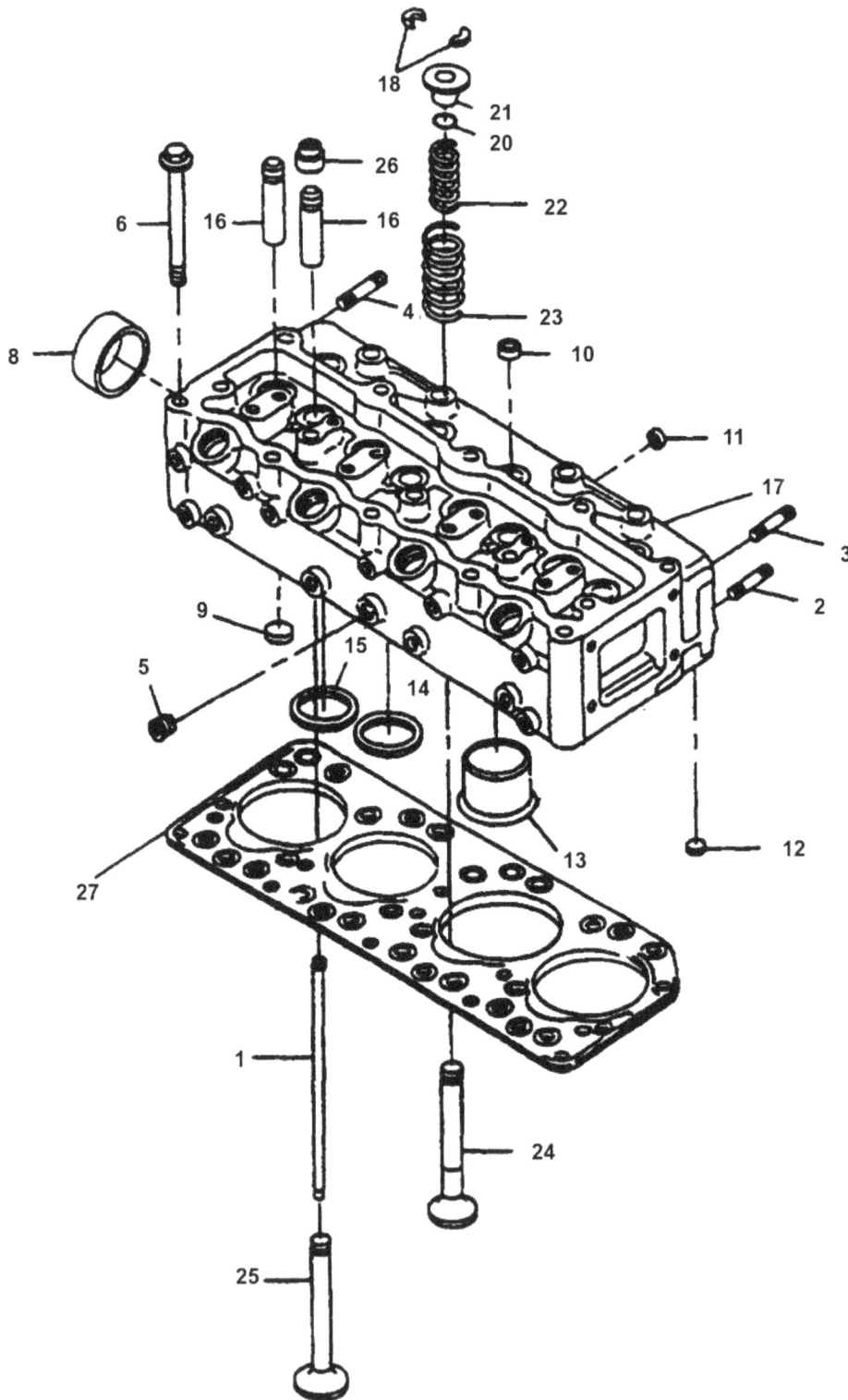

Figure 15. Cylinder Head Assembly.

(1)	(2)		(3)	(4)	(5)	(6)	(7)
	SMR CODE						
	a.	b.					
ITEM NO	**ARMY**	**AIR FORCE**	**NSN CAGEC PART NUMBER DESCRIPTION AND USABLE ON CODE (UOC)**				**QTY**

GROUP 06 CYLINDER HEAD
AND CRANKCASE

FIG. 15 CYLINDER HEAD
ASSEMBLY

1 PAFZZ PAOZZ 2815-01-221-7484 61888 Z5125750160 . PUSH ROD, ENGINE POP 8
2 PAFZZ PAOZZ 5307-01-226-8708 04627 324125 . STUD, PLAIN 3
3 PAFZZ PAOZZ 5307-01-226-4378 30076 324126 . STUD 3
4 PAFZZ PAOZZ 5307-01-226-4379 S0226 9041108450 . STUD 1
5 PAFZZ PAOZZ 4730-01-269-2313 S0226 5096050050 . PLUG , PIPE 1
6 PAFZZ PAOZZ 5306-01-265-2059 04627 1334663 . BOLT, MACHINE .... 19 7 PAFFH PAOOO 2815-01-227-6678 04627 323814 . CYLINDER HEAD, DIESEL "NOT

ILLUSTRATED" 1
8 PAFZZ PAOZZ 5340-01-269-5243 04627 323615 . . PLUG, EXPANSION RIGHT REAR,

DIA 44 MM 1
9 PAFZZ PAOZZ 5340-01-265-7264 04627 323816 . . PLUG, EXPANSION LOWER, DIA 16

MM 4
10 PAFZZ PAOZZ 5340-01-265-7263 04627 323817 . . PLUG, EXPANSION UPPER, DIA 18

MM 4
11 PAFZZ PAOZZ 5340-01-266-5116 04627 323818 . . PLUG, EXPANSION DIA 12MM 4
12 PAFZZ PAOZZ 5340-01-265-7271 04627 323819 . . PLUG, PROTECTIVE, DUS 4
13 PBFZZ PBOZZ 5340-01-226-4376 04627 323824 . . PLUG 4
14 PAFZZ PAOZZ 2815-01-222-0793 61888 Z5117150230 . . INSERT, ENGINE VALVE INTAKE .... 4 15 PAFZZ PAOZZ 2815-01-222-4317 04627 323821 . . INSERT, ENGINE VALVE EXHAUST,

PART OF KIT P/N 5878101862-2 4
16 PAFZZ PAOZZ 2815-01-222-1001 04627 323822 . . GUIDE, ENGINE POPPET 8
17 XAFZZ XA--- S0226 5111112000 . . HEAD, CYLINDER 1
18 PAFZZ PAOZZ 2815-01-221-7485 61688 Z5125650040 . LOCK, VALVE SPRING R .... 16 19 PAFZZ PAOZZ 2815-01-222-4318 61688 Z5125630160 . INSERT, ENGINE VALVE "NOT ILLUS-

TRATED" 8
20 KFFZZ KFF-- 5330-01-421-3133 04627 1369039 . . SEAL, PLAIN, PART OF KIT P/N 587810-9401, KIT P/N 5878101862-2 1
21 XAFZZ XA--- S0266 5125630170 . . SEAT, ASSEMBLY 1
22 PAFZZ PAOZZ 5360-01-250-0562 S0226 5-12562007-0 . SPRING, HELICAL, COMP 8
23 PAFZZ PAOZZ 5360-01-226-6674 04627 324303 . SPRING, HELICAL, COMP .... 8 24 PAFZZ PAOZZ 2815-01-362-5668 S0226 512552-0340 . VALVE, POPPET ENGINE

EXHAUST 4
25 PAFZZ PAOZZ 2815-01-363-5159 S0226 5125510282 . VALVE, POPPET ENGINE INTAKE ..... 4 26 PAFZZ PAOZZ 5330-01-226-4377 04627 323823 . SEAL, PART OF KIT P/N 587810-9401,

KIT P/N 5878101862-2 8
27 KFFZZ KFF-- 5330-01-226-4451 04627 1552967 . GASKET, CYL, HEAD, PART OF KIT P/N 567810-9401, KIT P/N 5878101862-2 1
28 PAFZZ PAOZZ 2815-01-381-9068 S0226 5878127130 . REPAIR KIT, DIESEL E 1

GASKET (Figure 1, Item 18)............. 1
GASKET (Figure 9, Item 11)............. 1
GASKET (Figure 14, Item 3)............. 2 GASKET, CYL HEAD (Figure 15, Item 27)............................. 1 INSERT, ENGINE V (Figure 15, Item 15)............................. 1

(1)	(2)		(3)	(4)	(5)	(6)	(7)
	SMR CODE						
	a.	b.					
ITEM NO	**ARMY**	**AIR FORCE**				**NSN CAGEC PART NUMBER DESCRIPTION AND USABLE ON CODE (UOC)**	**QTY**

O-RING (Figure 14, Item 6) 1
SEAL (Figure 3, Item 2) 2
SEAL (Figure 3, Item 12)................. 2
SEAL (Figure 15, Item 26)................ 1
SEAL, PLAIN (Figure 15, Item 20) 1
WASHER (Figure 3, Item 25)............ 4
WASHER (Figure 3, Item 26)............ 4

END OF FIGURE

FIELD AND SUSTAINMENT MAINTENANCE

DIESEL ENGINE, FOUR-CYLINDER, FOUR CYCLE, FUEL INJECTED C-240PW-28

GROUP 06 CYLINDER HEAD AND CRANKCASE: PISTONS, CONNECTING RODS, AND CRANKSHAFT

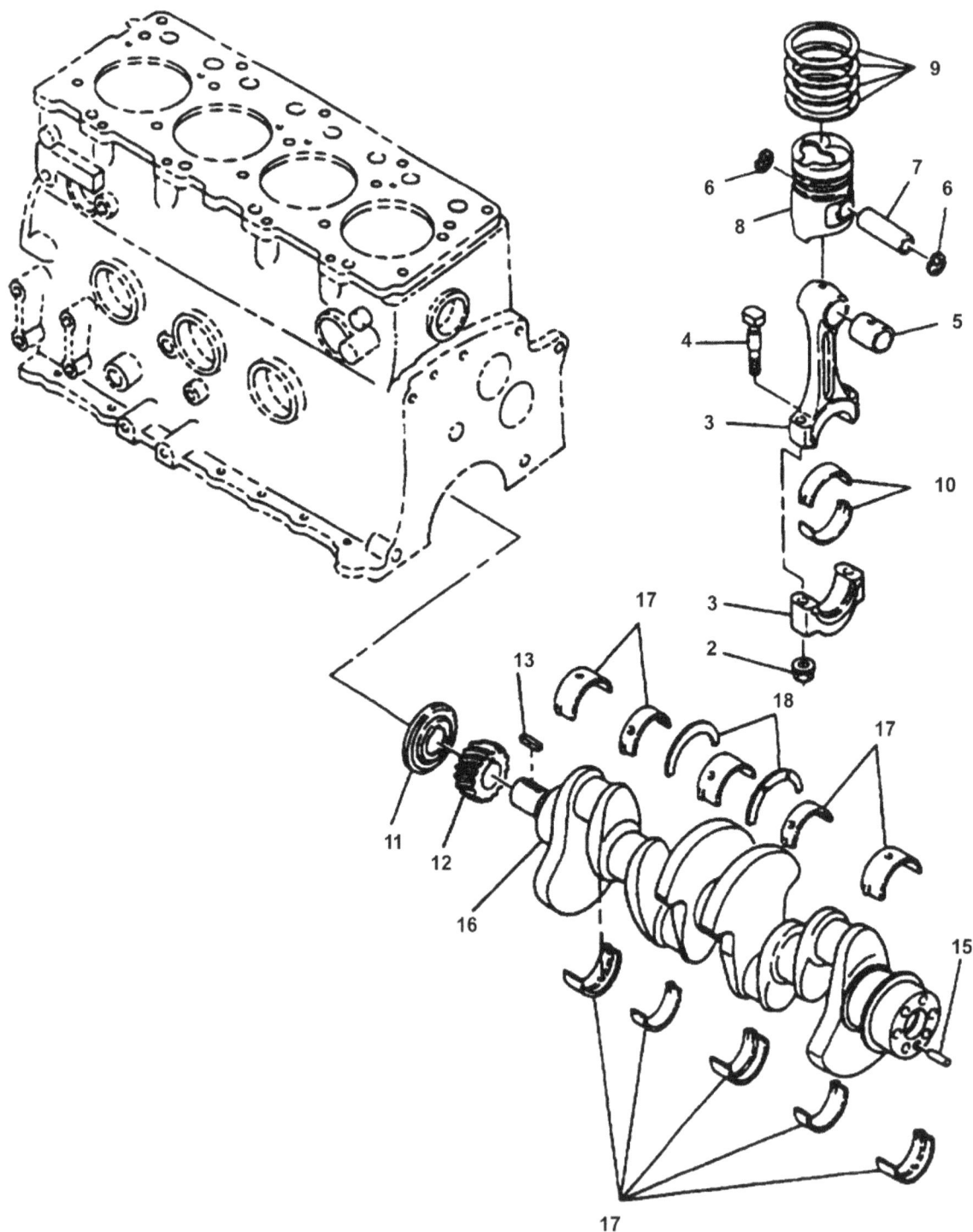

Figure 16. Pistons, Connecting Rods, and Crankshaft.

(1)	(2)		(3)	(4)	(5)	(6)	(7)
	SMR CODE						
	a.	b.					
ITEM NO	ARMY	AIR FORCE	NSN CAGEC PART NUMBER DESCRIPTION AND USABLE ON CODE (UOC)				QTY

GROUP 06 CYLINDER HEAD
AND CRANKCASE

FIG. 16 PISTONS, CONNECTING
RODS, AND CRANKSHAFT

1 PAHZZ PAOZZ 2815-01-360-6518 30076 5122300391 . CONNECTING ROD, PIST "NOT
ILLUSTRATED" 4

2 PAHZZ PAOZZ 5310-01-224-9101 04627 348457 . . NUT 2

3 XAHZZ XA--- S0226 512230-0391-1 . . CAP, ROD, BEARING 1

4 PAHZZ PAOZZ 5306-01-223-6484 04627 348456 . . BOLT .... 2 5 PAHZZ PAOZZ 3120-01-226-6701 04627 348455 . . BEARING, SLEEVE, PART OF KIT P/
N 5878102361 1

6 PAHZZ PAOZZ 5365-01-226-6698 04627 324271 . RING, RETAINING, PART OF KIT P/ N
5878102361 8

7 PAHZZ PAOZZ 2815-01-222-0250 04627 324270 . PIN, PISTON, PART OF KIT P/N
5878102361 4

8 PAHZZ PAOZZ 2815-01-222-0403 30076 324268 . PISTON, INTERNAL COM, PART OF
KIT P/N 5878102361 4

9 PAHZZ PAOZZ 2815-01-248-6340 S0226 512121-0070 . RING SET, PISTON, PART OF KIT P/N
5878102361 4

10 PAHZZ PAOZZ 3120-01-224-8104 04627 362120 . BEARING SET, SLEEVE
STANDARD 4

10 PAHZZ PAOZZ 3120-01-226-6706 04627 362121 . BEARING HALF SET, SL .25 MM
UNDERSIZE 4

10 PAHZZ PAOZZ 3120-01-223-7073 04627 362122 . BEARING SET, SLEEVE SO MM
UNDERSIZE 4

11 PBHZZ PBOZZ 2815-01-222-0266 30076 324285 . DEFLECTOR, DIRT AND 1

12 XDHZZ XB--- 3020-01-222-3826 04627 324284 . GEAR, HELICAL 1

13 PAHZZ PAOZZ 5315-01-361-5705 S0226 9080307160 . KEY, WOODRUFF .... 1 14 PAHHH PAHHH 2815-01-363-2382 4N530 8941421920 . CRANKSHAFT, ENGINE "NOT ILLUS-
TRATED" 1

15 PAHZZ PAOZZ 5340-01-269-2709 04627 324277 . . DISK, SOLID, PLAIN 1

16 XAHZZ XA--- 4N530 8941421930 . . CRANKSHAFT .... 1 17 PAHZZ PAOZZ 3120-01-228-1465 04627 324265 . BEARING HALF SET, SL .25 MM
UNDERSIZE 1

17 PAHZZ PAOZZ 3120-01-224-8103 04627 324266 . BEARING HALF SET, SL .50 MM
UNDERSIZE 1

17 PAHZZ PAOZZ 3120-01-250-3235 S0226 511510-0211 . BEARING SET, SLEEVE
STANDARD 1

18 PAHZZ PAOZZ 3120-01-365-1738 30076 133003 . BEARING, WASHER, THRU 2

END OF FIGURE

FIELD AND SUSTAINMENT MAINTENANCE

DIESEL ENGINE, FOUR-CYLINDER, FOUR CYCLE, FUEL INJECTED C-240PW-28
GROUP 06 CYLINDER HEAD AND CRANKCASE: CAMSHAFT ASSEMBLY

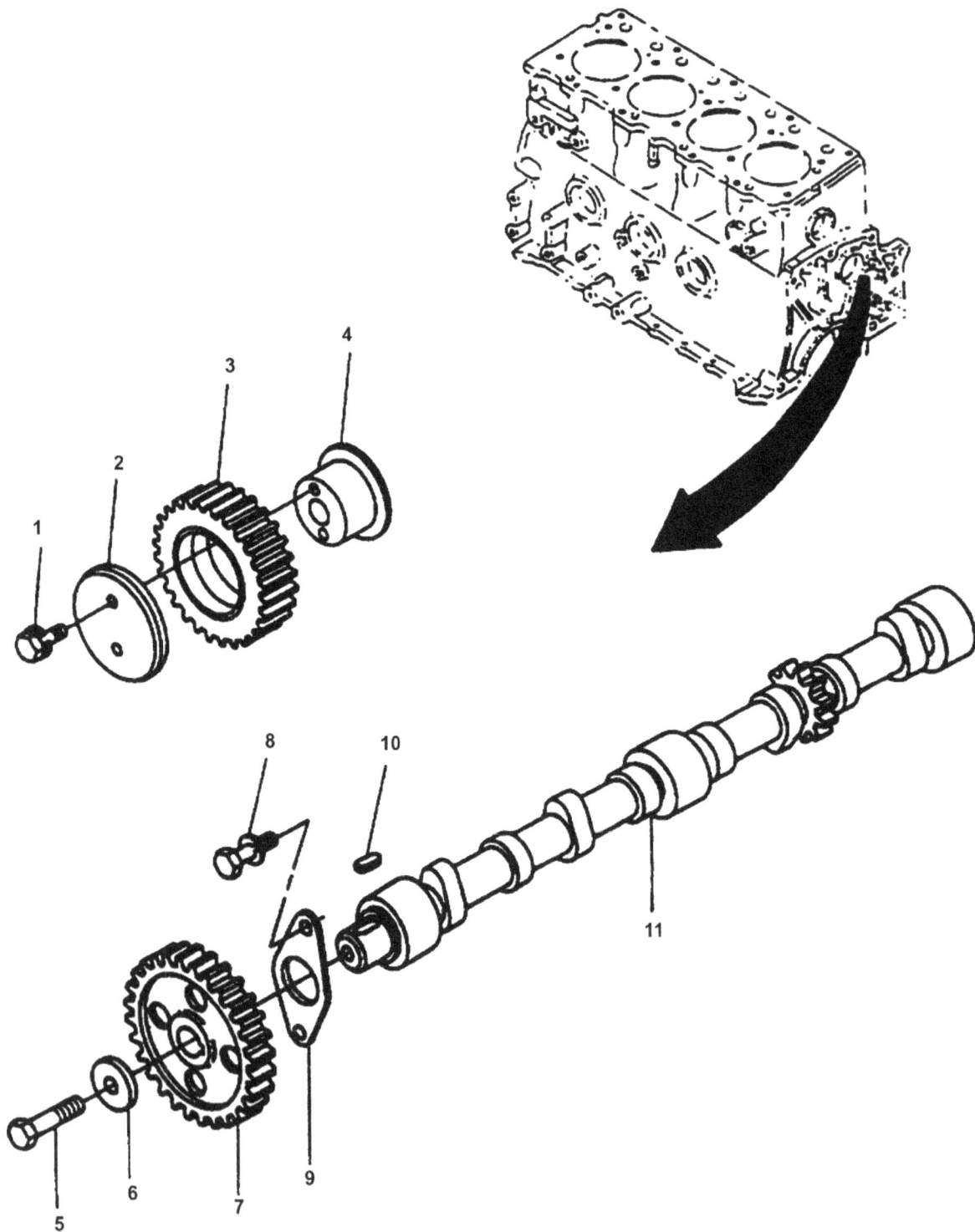

Figure 17. Camshaft Assembly.

(1)	(2)	(3)	(4)	(5)	(6)	(7)
	SMR CODE					
	a. b.					
ITEM NO	**ARMY AIR FORCE**		NSN CAGEC PART NUMBER DESCRIPTION AND USABLE ON CODE (UOC)			**QTY**

GROUP 06 CYLINDER HEAD
AND CRANKCASE

FIG. 17 CAMSHAFT ASSEMBLY

```
 1 PAHZZ PAOZZ 5306-01-361-6941 4F040 894150-045 . BOLT, MACHINE ......................... 2
 2 PBHZZ PBOZZ 3010-01-263-7188 04627 324299 . THRUST PLATE, TRANS ............... 1
 3 PBHZZ PBOZZ 3020-01-363-3091 04627 8-97166-030-0 . GEAR, HELICAL .......................... 1
 4 PAHZZ PAOZZ 2590-01-262-2677 30076 324297 . ROD, PISTON, LINEAR A ............... 1
 5 PBHZZ XB--- 5306-01-265-2061 S0226 5090000690 . BOLT, MACHINE ......................... 1
 6 PAHZZ PAOZZ 5310-01-265-6689 30076 324294 . WASHER, FLAT .......................... 1
 7 PBHZZ PBOZZ 3020-01-222-4632 30076 324291 . GEAR, HELICAL .......................... 1
 8 PAHZZ PAOZZ 5305-01-267-9281 04627 1338410 . SCREW, ASSEMBLED WAS ........... 2
 9 XDHZZ XB--- S0226 9116810101 . PLATE, THRUST, CAM .................. 1
10 PAHZZ PAOZZ 5315-01-226-8671 04627 324292 . KEY ........................................ 1
11 PAHZZ PAOZZ 2815-01-406-1958 S0226 5125110784 . CAMSHAFT, ENGINE ................... 1
```

END OF FIGURE

FIELD AND SUSTAINMENT MAINTENANCE

DIESEL ENGINE, FOUR-CYLINDER, FOUR CYCLE, FUEL INJECTED C-240PW-28

GROUP 06 CYLINDER HEAD AND CRANKCASE: FLYWHEEL AND HOUSING

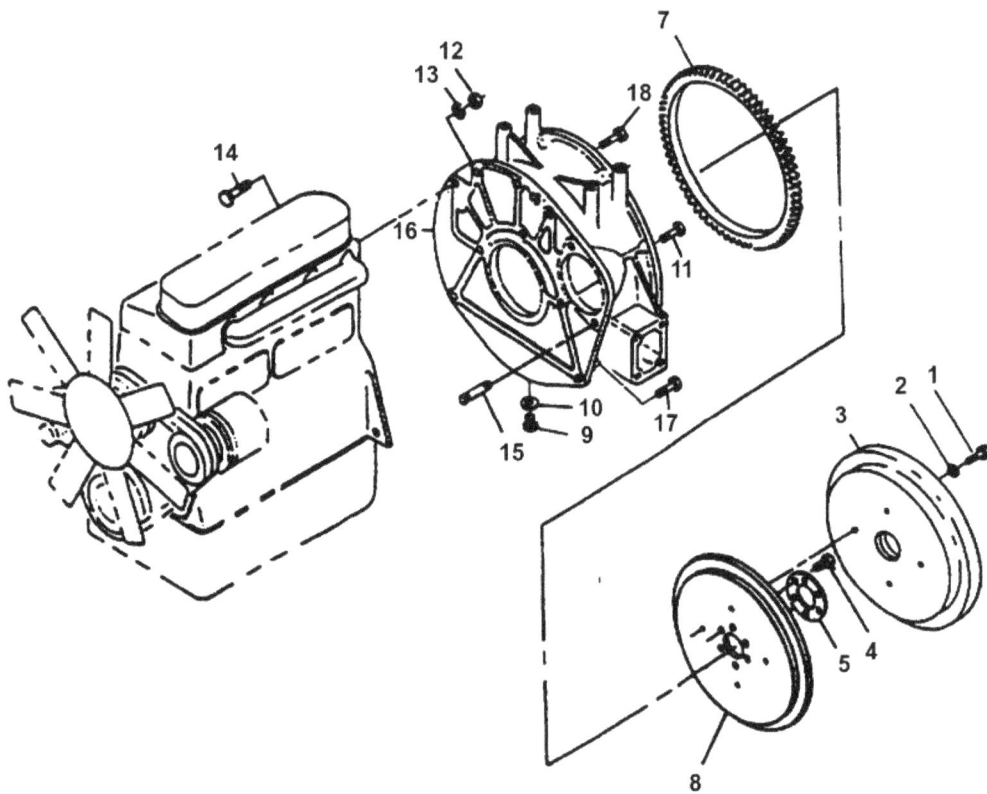

Figure 18. Flywheel and Housing.

(1)	(2)		(3)	(4)	(5)	(6)	(7)
	SMR CODE						
	a.	b.					
ITEM NO	ARMY	AIR FORCE	NSN CAGEC PART NUMBER DESCRIPTION AND USABLE ON CODE (UOC)				QTY

GROUP 06 CYLINDER HEAD
AND CRANKCASE

FIG. 18 FLYWHEEL AND
HOUSING

1 PAFZZ PAOZZ 5306-01-467-5689 S0226 8-97109-673-0 . BOLT, MACHINE 4
2 PAFZZ PAOZZ 5310-01-361-5719 S0226 9098518660 . WASHER, FLAT 4
3 XDFZZ XB--- 4N530 5123310500 . FLY WHEEL 1
4 PAFZZ PAOZZ 5306-01-361-6606 4F040 8944339910 . BOLT, MACHINE 6
5 XDFZZ XB--- 4N530 5123366101 . WASHER, FLY WHEEL .... 1 6 XDFFF XB--- 4N530 5123301080 . FLY WHEEL ASSEMBLY "NOT ILLUS-

TRATED" 1
7 PBFZZ PBOZZ 3020-01-362-5712 9M091 9123330361 . . GEAR, SPUR 1
8 XAFZZ XA--- 4N530 512330-1080-1 . . FLYWHEEL, BASIC 1
9 XDFZZ XB---   5365-01-265-7189 04627 348427 . PLUG, MACHINE 1
10 XDFZZ XB--- 5339-01-382-5916 S0226 9095714160 . GASKET 1
11 PAFZZ PAOZZ 5306-01-265-2060 S0226 5090002270 . BOLT, MACHINE 3
12 PAFZZ PAOZZ 5310-01-254-6368 45152 401.034 . NUT, PLAIN, HEXAGON 1
13 PAFZZ PAOZZ 5310-01-361-5722 S0226 9091505100 . WASHER, LOCK 1
14 PAFZZ PAOZZ 5306-01-384-3476 4F040 9096065350 . BOLT, MACHINE 1
15 PAFZZ PAOZZ 5307-01-265-7140 30076 324222 . STUD, PLAIN 2
16 XDFZZ XDOZZ 4N530 8970698070 . HOUSING, FLY WHEEL 1
17 PAFZZ PAOZZ 5306-01-467-4944 S0226 500412320 . BOLT 2
18 PAFZZ PAOZZ 5306-01-467-5694 S0226 0-28081-030-0 . BOLT 3

END OF FIGURE

FIELD AND SUSTAINMENT MAINTENANCE

DIESEL ENGINE, FOUR-CYLINDER, FOUR CYCLE, FUEL INJECTED C-240PW-28

GROUP 06 CYLINDER HEAD AND CRANKCASE: CYLINDER BLOCK ASSEMBLY

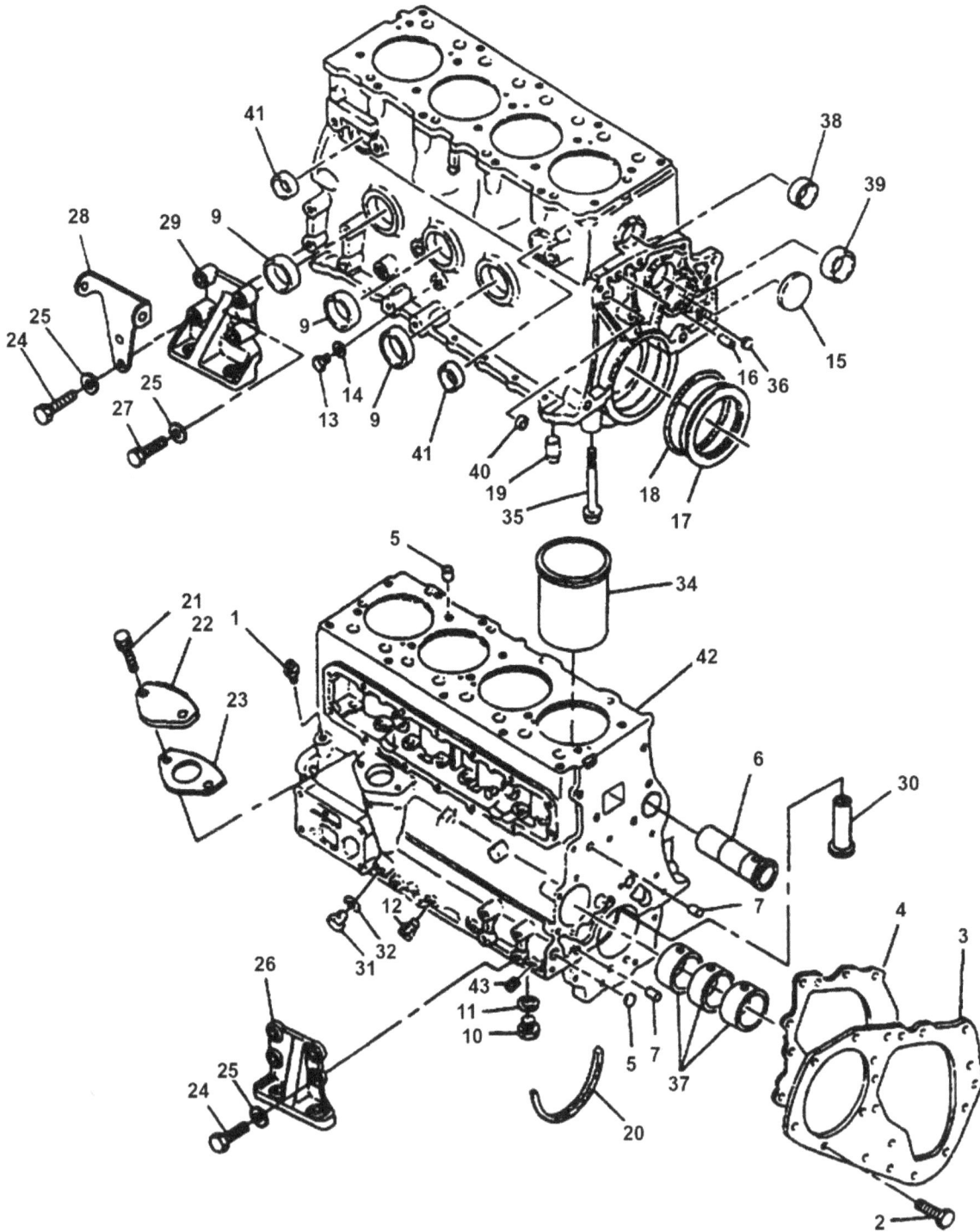

Figure 19. Cylinder Block Assembly.

(1) ITEM NO	(2) SMR CODE a.　b. ARMY　AIR 　　　FORCE	(3) NSN	(4) CAGEC	(5) PART NUMBER	(6) DESCRIPTION AND USABLE ON CODE (UOC)	(7) QTY

GROUP 06 CYLINDER HEAD
AND CRANKCASE

FIG. 19 CYLINDER BLOCK
ASSEMBLY

```
  1 PAFZZ PAOZZ 4730-01-221-8816 04627 324154 . FITTING ................................... 1
  2 PAHZZ PAOZZ 5306-01-268-6796 30076 324249 . BOLT, MACHINE ........................ 3
  3 XDHZZ XB--- 4N530 9113111530 . PLATE, FRT., CYLINDER ............... 1 4 PAHZZ PAOZZ 5330-01-226-4450 30076 324152 .
GASKET, PART OF KIT P/N
                                                         587810-9401 .............................. 1
  5 XDFZZ XB--- 4N530 8944533440 . DOWEL, CYLINDER BLO ............... 2
  6 XDHZZ XB--- 4N530 8944548320 . DUCT, WATER, CYLIND ............... 1
  7 XDHZZ XB--Z 4N530 9081510240 . PIN, STRAIGHT, HEAD ................. 2
  8 PAHZZ PAOZZ 5340-01-265-7264 04627 323816 . PLUG EXPANSION ...................... 2
  9 PAHZZ PAOZZ 5340-01-269-5243 04627 323815 . PLUG, EXPANSION ..................... 3
 10 PAFZZ PAOZZ 5365-01-265-7189 04627 348427 . PLUG, MACHINE THREAD ............. 1
 11 PAFZZ PAOZZ 5339-01-382-5916 S0226 9095714160 . GASKET .................................... 1
 12 PAFZZ PAOZZ 4730-01-362-5574 04627 1352025 . NIPPLE, PIPE ............................ 1
 13 PAFZZ PAOZZ 5365-01-361-5708 S0226 9092023080 . PLUG, MACHINE THREAD ............. 1 14 PAFZZ PAOZZ 5330-01-224-9108 04627 324412 . SEAL, PART OF KIT P/N
                                                         587810-9401 .............................. 1
 15 PAHZZ PAOZZ 5340-01-361-8366 S0226 8942505140 . PLUG, EXPANSION ..................... 1
 16 XDHZZ XB--- 4N530 9081610320 . PIN, STRAIGHT, HEAD ................. 1 17 PAFZZ PAOZZ 5330-01-361-6958 S0226
8941117500 . SEAL, PLAIN, PART OF KIT P/N
                                                         587810-9401 .............................. 1
 18 XDFZZ XB--- 4N530 9123631121 . SPACER, OIL SEAL ...................... 1 19 XDHZZ XDHZZ 4N530 9098785970 . PIN, DOWEL,
PART OF KIT P/N
                                                         587810-9401 .............................. 2
 20 PAHZZ PAOZZ 5330-01-224-9098 04627 324165 . SEAL, PART OF KIT P/N
                                                         587810-9401 .............................. 2
 21 PAFZZ PAOZZ 5305-01-270-1704 04627 324171 . SCREW, ASSEMBLED WAS ........... 2
 22 XDFZZ XB--- 4N530 9112190930 . COVER, OIL PUMP ...................... 1 23 PAFZZ PAOZZ 5330-01-226-4448 04627 324157 .
GASKET, PART OF KIT P/N
                                                         587810-9401 .............................. 1
 24 PAFZZ PAOZZ 5306-01-388-2096 4F040 026091-025 . BOLT, MACHINE ......................... 5
 25 PAFZZ PAOZZ 5310-01-361-5719 S0226 9098518660 . WASHER, FLAT ......................... 8
 26 XDFZZ XB--- 4N530 5117710220 . BRACKET, ANGLE ...................... 1
 27 PAFZZ PAOZZ 5306-01-361-4588 4N530 0-28061-040-0 . BOLT, MACHINE ......................... 3
 28 PBFZZ PBOZZ 5340-01-414-2903 1JY58 5197110850 . BRACKET, MOUNTING ................. 1
 29 XDFZZ XB--- 4N530 5117711500 . FOOT, ENGINE, LH ..................... 1
 30 PAHZZ PAOZZ 2815-01-222-7733 0LUY9 324307 . TAPPET, ENGINE POPPET ............ 8
 31 PAFZZ PAOZZ 5365-01-416-6508 4N530 9992023140 . PLUG, MACHINE ....................... 1
 32 PAFZZ PAOZZ 5330-01-414-3713 4N530 894158328 . GASKET .................................. 1 33 PBHHH PBOOO 2815-01-362-
3546 S0226 5112101774 . ENGINE BLOCK ASSEMBLY "NOT
                                                         ILLUSTRATED" .......................... 1
 34 PAHZZ PAOZZ 2815-01-222-0402 30076 324142 .. CYLINDER SLEEVE, PART OF KIT P/
                                                         N 5878102361 ........................... 4
 35 PAHZZ PAOZZ 5306-01-265-4697 S0226 9098025910 .. BOLT, MACHINE ...................... 10
 36 PAHZZ PAOZZ 5340-01-265-7264 04627 323816 .. PLUG, EXPANSION ................... 2
 37 PAHZZ PAOZZ 3120-01-226-6801 04627 324143 .. BEARING SET, SLEEVE ............. 1
 38 PAFZZ PAOZZ 5340-01-265-6749 04627 324139 .. PLUG, EXPANSION ................... 1
```

(1)	(2) SMR CODE		(3)	(4)	(5)	(6)	(7)
	a.	b.					
ITEM NO	ARMY	AIR FORCE	NSN	CAGEC	PART NUMBER	DESCRIPTION AND USABLE ON CODE (UOC)	QTY

39 PAFZZ PAOZZ 5340-01-266-5117 04627 324140 . . PLUG, EXPANSION 1
40 PAFZZ PAOZZ 5340-01-266-5116 04627 323818 . . PLUG, EXPANSION 1
41 PAFZZ PAOZZ 5340-01-265-6748 04627 1456807 . . PLUG, EXPANSION 2
42 XAHZZ XAHZZ 4N530 511210-1773-1 . . BLOCK, BASIC 1
43 PAHZZ PAOZZ 4730-01-363-4324 S0226 509605-019 . . PLUG, PIPE 1
44 PAHZZ PAOZZ   2815-389-0315 S0226 5878102361 . CYLINDER SLEEVE KIT 4
                                     S0226 587814-5810 . CYLINDER SLEEVE GRADE 1 1
                                     S0226 587814-5820 . CYLINDER SLEEVE GRADE 2 1
                                     S0226 587814-5830 . CYLINDER SLEEVE GRADE 3 1
                                     S0226 587814-5840 . CYLINDER SLEEVE GRADE 4 1
                                                       BEARING, SLEEVE (Figure 16, Item 5) .... 1 CYLINDER SLEEVE (Figure 19, Item 34)... 4
                                                       PIN, PISTON (Figure 16, Item 7)........ 4
                                                       PISTON (Figure 16, Item 8)............... 4
                                                       RING, RETAIN (Figure 16, Item 6)....... 8 RING SET, PISTON (Figure 16, Item 9) 4
45 PAHZZ PAOZZ   587810-9401 . REPAIR KIT, DIESEL E 1
                                                       GASKET (Figure 1, Item 7) 1
                                                       GASKET (Figure 1, Item 18).............. 8
                                                       GASKET (Figure 4, Item 2) 1
                                                       GASKET (Figure 5, Item 2) 2
                                                       GASKET (Figure 5, Item 5) 2
                                                       GASKET (Figure 6, Item 10)............. 1
                                                       GASKET (Figure 8, Item 35)............. 1
                                                       GASKET (Figure 9, Item 9) 1
                                                       GASKET (Figure 9, Item 11)............. 1
                                                       GASKET (Figure 14, Item 3)............. 2
                                                       GASKET (Figure 19, Item 4)............. 1
                                                       GASKET (Figure 9, Item 23)............. 1 GASKET, CYL HEAD (Figure 15, Item 27)... 1
                                                       O-RING (Figure 4, Item 70) 2
                                                       O-RING (Figure 5, Item 20) 1
                                                       O-RING (Figure 13, Item 9) 1
                                                       O-RING (Figure 14, Item 6) 1
                                                       PIN, DOWEL (Figure 19, Item 19)........ 1
                                                       SEAL (Figure 2, Item 2) 2
                                                       SEAL (Figure 2, Item 7) 2
                                                       SEAL (Figure 3, Item 2) 2
                                                       SEAL (Figure 3, Item 12)................... 2
                                                       SEAL (Figure 15, Item 26)................. 8
                                                       SEAL (Figure 19, Item 14)................. 1
                                                       SEAL (Figure 19, Item 17)................. 2
                                                       SEAL (Figure 19, Item 20)................. 1
                                                       SEAL, PLAIN (Figure 15, Item 20) 1
                                                       SEAL, PLAIN (Figure 13, Item 3)......... 1
                                                       SEAL, PLAIN (Figure 13, Item 7)......... 1
                                                       WASHER (Figure 3, Item 25)............. 4

(1)	(2)		(3)	(4)	(5)	(6)	(7)
	SMR CODE						
	a.	b.					
ITEM NO	ARMY	AIR FORCE	NSN CAGEC PART NUMBER DESCRIPTION AND USABLE ON CODE (UOC)				QTY

WASHER (Figure 3, Item 26)............. 4

END OF FIGURE

FIELD AND SUSTAINMENT MAINTENANCE

DIESEL ENGINE, FOUR-CYLINDER, FOUR CYCLE, FUEL INJECTED C-240PW-28
GROUP 07 BULK ITEMS

(1)	(2)		(3)	(4)	(5)	(6)	(7)
	SMR CODE						
	a.	b.					
ITEM NO	ARMY	AIR FORCE	NSN	CAGEC	PART NUMBER	DESCRIPTION AND USABLE ON CODE (UOC)	QTY

GROUP 07 BULK ITEMS

1 PAFZZ PAOZZ 4720-01-302-2944 04627 1456806 . HOSE, NONMETALLIC 1
2 PAFZZ PAOZZ 4720-01-359-9633 81349 M6000-D-1440 . HOSE, NONMETALLIC 1

END OF FIGURE

FIELD AND SUSTAINMENT MAINTENANCE

DIESEL ENGINE, FOUR-CYLINDER, FOUR CYCLE, FUEL INJECTED C-240PW-28
SPECIAL TOOLS LIST

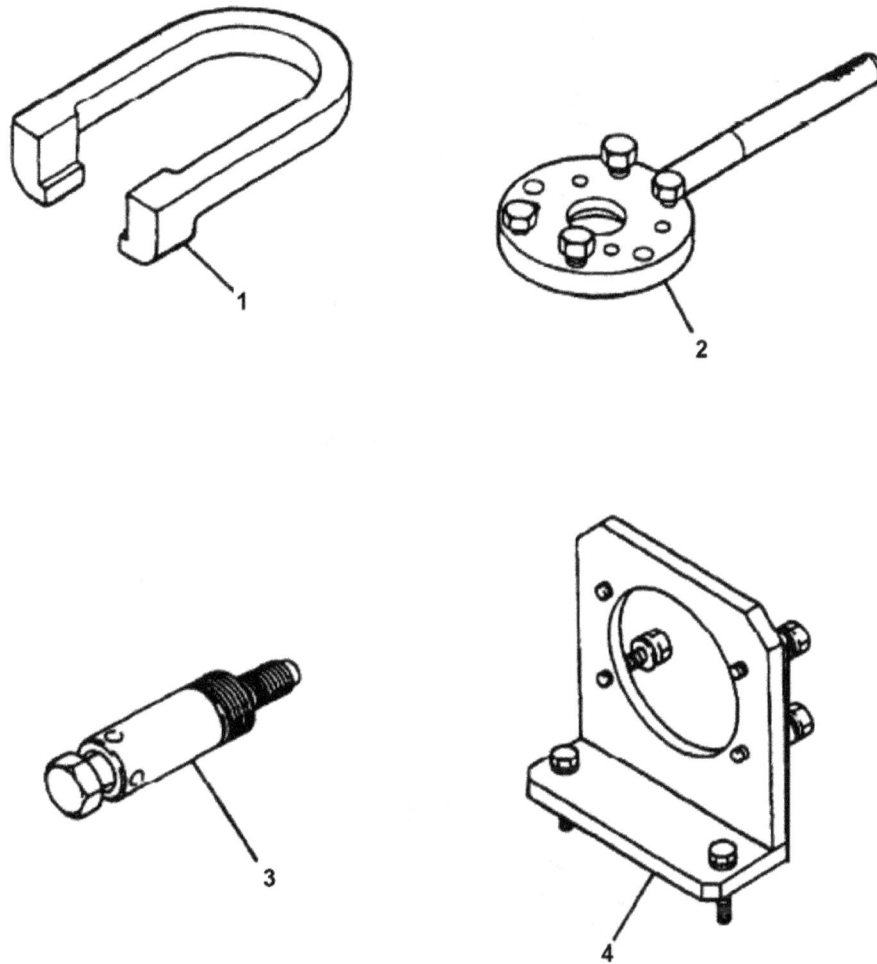

Figure 1. Special Tools (Sheet 1 of 4).

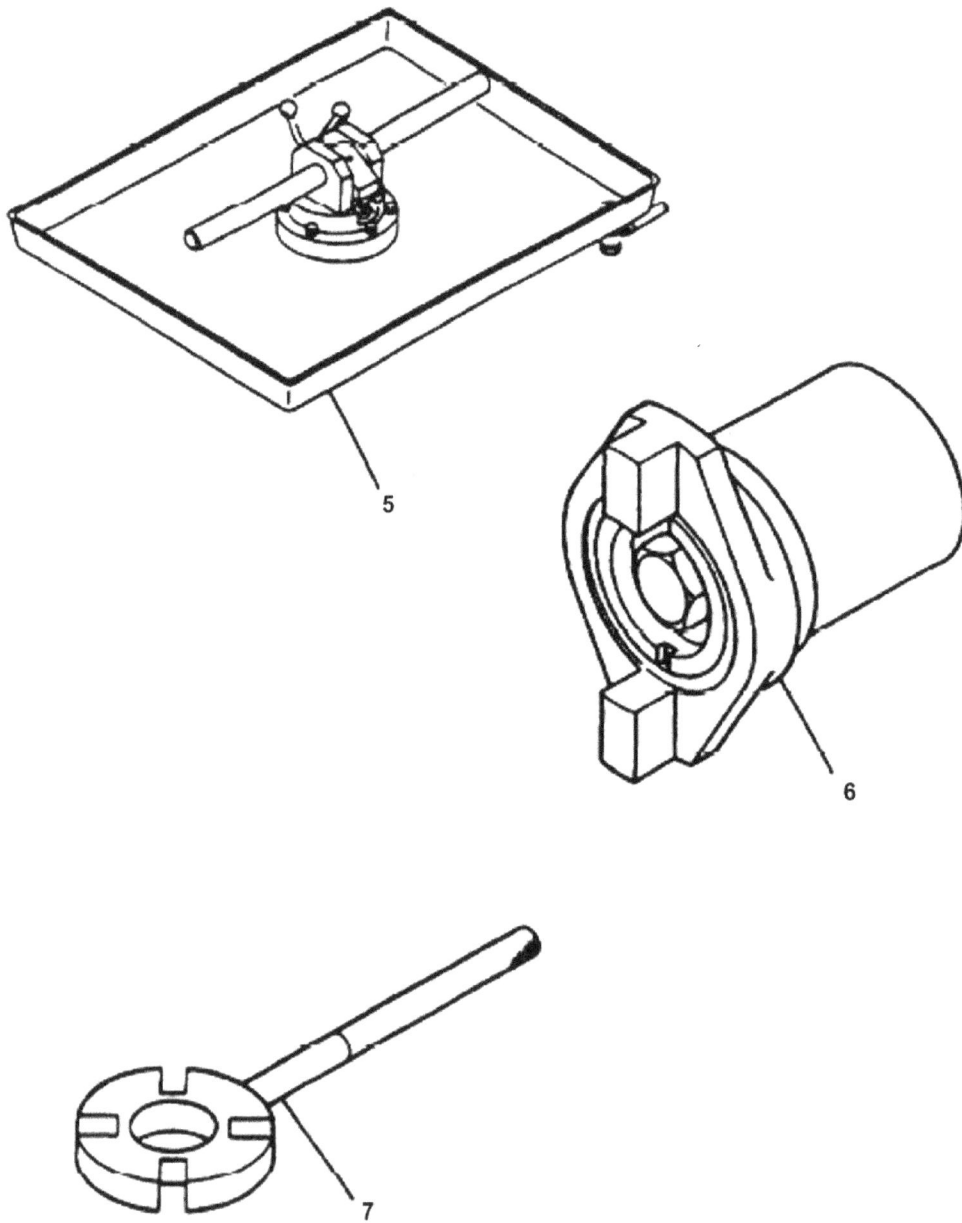

Figure 1. Special Tools (Sheet 2 of 4).

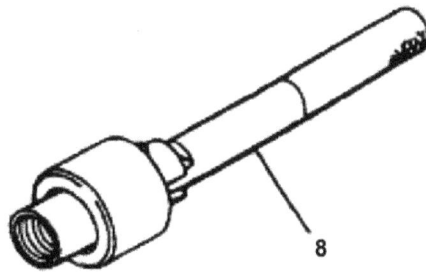

8

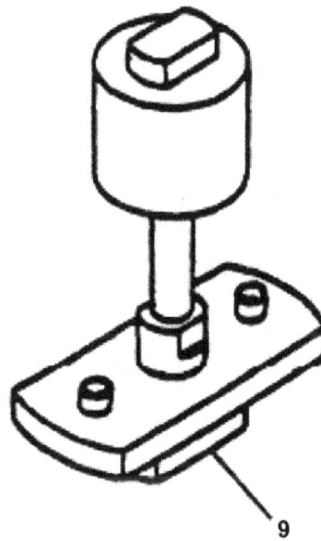

9

Figure 1. Special Tools (Sheet 3 of 4).

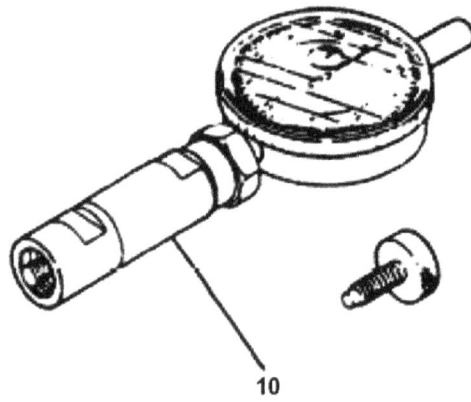

10

11

Figure 1. Special Tools (Sheet 4 of 4).

(1)	(2)		(3)	(4)	(5)	(6)	(7)
	SMR CODE						
	a.	b.					
ITEM NO	ARMY	AIR FORCE	NSN CAGEC PART NUMBER DESCRIPTION AND USABLE ON CODE (UOC)				QTY

GROUP 12 SPECIAL TOOLS

FIG. 1 SPECIAL TOOLS

1 PEHZZ PEOZZ 5120-01-390-8345 4N530 57931-4100 . PLUNGER, SPRING HOLD 1
2 PEHZZ PEOZZ 5120-01-390-7577 4N530 157916-5320 . WRENCH, SPANNER 1
3 PEHZZ PEOZZ 5120-01-390-8342 4N530 57931-4100 . EXTRACTOR, TIMING DE 1
4 PEHZZ PEOZZ 5120-01-390-8347 4N530 157944-7820 . BRACKET, PUMP REPAIR 1
5 PEHZZ PEOZZ 5120-01-390-7570 4N530 157944-6520 . VISE, MULTIPOSITION 1
6 PEHZZ PEOZZ 5120-01-390-8346 4N530 157842-4420 . COUPLING, INJECTION 1
7 PEHZZ PEOZZ 5120-01-390-7574 4N530 157916-5420 . WRENCH, SPANNER 1
8 PEHZZ PEOZZ 5120-01-390-8343 4N530 157920-9620 . EXTRACTOR 1
9 PEHZZ PEOZZ 5120-01-390-4740 4N530 157924-1620 . INSERTER, FLYWEIGHT 1
10 PEHZZ PEOZZ 5210-01-390-8341 4N530 105782-6280 . GAGE, CONTROL RACK 1
11 PEHZZ PEOZZ 5210-01-390-8344 4N530 105782-4020 . GAGE, TAPPET CLEAR .... 1 12 PEHZZ PEOZZ 5120-01-398-3999 4N530 5-8522-0013-0 . INSERTER, SEAL INSTALL CRANK-

SHAFT FRONT OIL SEAL 1

END OF FIGURE

END OF WORK PACKAGE

FIELD AND SUSTAINMENT MAINTENANCE

DIESEL ENGINE, FOUR-CYLINDER, FOUR CYCLE, FUEL INJECTED C-240PW-28
NATIONAL STOCK NUMBER (NSN) INDEX

STOCK NUMBER	FIG.	ITEM	STOCK NUMBER	FIG.	ITEM
3110-00-144-8499	12	35	5310-01-224-8007	14	10
5306-01-102-2644	14	7	5330-01-224-8063	4 49	
5310-01-168-8140	12	12	5330-01-224-8068	6 10	
2910-01-217-3645	3 24		5330-01-224-8070	9 11	
6620-01-220-7105	1 13		5330-01-224-8071	9	9
4710-01-221-2214	5	3	3120-01-224-8103	16	17
4710-01-221-5770	3	3	3120-01-224-8104	16	10
4730-01-221-7154	2	8	5330-01-224-9098	19	20
2805-01-221-7364	3 27		5306-01-224-9099	1	15
2815-01-221-7484	15	1		12	1
2815-01-221-7485	15	18	5330-01-224-9100	2	2
4730-01-221-8816	19	1		2	7
2815-01-221-8925	14	17	5310-01-224-9101	16	2
2815-01-221-8926	14	16	5360-01-224-9102	14	15
2815-01-222-0250	16	7	5310-01-224-9104	9	2
2815-01-222-0266	16	11	5310-01-224-9107	3 26	
2815-01-222-0402	19	34	5330-01-224-9108	3	2
2815-01-222-0403	16	8		3	12
2815-01-222-0793	15	14		19	14
2815-01-222-1001	15	16	5365-01-224-9109	4 66	
3040-01-222-3752	14	20	5340-01-226-4376	15	13
3020-01-222-3826	16	12	5330-01-226-4377	15	26
2815-01-222-4317	15	15	5307-01-226-4378	15	3
2815-01-222-4318	15	19	5307-01-226-4379	15	4
3020-01-222-4632	17	7	5330-01-226-4448	19	23
3040-01-222-4717	14	13	5330-01-226-4450	19	4
2815-01-222-5491	14	14	5330-01-226-4451	15	27
2815-01-222-5492	14	12	5330-01-226-4454	1 18	
4730-01-222-7562	3	1	5330-01-226-4455	14	3
2815-01-222-7733	19	30	5310-01-226-6660	3 28	
5306-01-223-6484	16	4	5310-01-226-6661	3 25	
3120-01-223-7073	16	10	5330-01-226-6672	13	7
			5360-01-226-6674	15	23

STOCK NUMBER	FIG.	ITEM	STOCK NUMBER	FIG.	ITEM
5330-01-226-6692	5	5	3120-01-250-3235	16	17
5330-01-226-6695	1	7	2910-01-251-2499	4 11	
5365-01-226-6698	16	6	3110-01-252-0446	12	30
3120-01-226-6701	16	5	5310-01-254-6368	18	12
3120-01-226-6706	16	10	2590-01-260-0382	14	5
3120-01-226-6801	19	37	2910-01-261-0112	4	6
5315-01-226-8671	17	10	4720-01-261-4182	8 20	
3110-01-226-8680	4 64		4730-01-261-6818	8 18	
3110-01-226-8681	4 64		2510-01-262-2650	9 15	
5330-01-226-8686	4 20		2990-01-262-2653	10	1
5330-01-226-8690	4 73		2590-01-262-2677	17	4
5330-01-226-8691	4 27		2815-01-262-2760	9	7
5330-01-226-8700	4 52		4720-01-262-6271	8	5
5307-01-226-8708	15	2	2590-01-263-3149	8 27	
5365-01-226-8764	4 65		2815-01-263-3150	6 17	
5365-01-226-8766	4 65		2930-01-263-3162	1 17	
5306-01-226-9105	5	4	2990-01-263-3171	2 23	
5365-01-227-0612	4 65		2815-01-263-3179	8 34	
5365-01-227-0633	4 65		2805-01-263-6472	8 32	
5340-01-227-3183	8 17		4730-01-263-7175	8	4
2815-01-227-6678	15	7	3010-01-263-7188	17	2
5365-01-227-9663	4 65		2990-01-263-7230	8 29	
5365-01-227-9883	4 65		4730-01-264-3345	4 72	
3120-01-228-1465	16	17	2530-01-264-3410	4 56	
4820-01-232-5701	5 15		2990-01-264-3412	4 48	
5310-01-232-8491	12	14	2805-01-264-3498	4 84	
2815-01-248-6340	16	9	4730-01-264-3880	8	6
2910-01-248-6547	3 24		4820-01-264-5571	4 10	
4710-01-248-8113	3	5	2910-01-264-5626	4 77	
4710-01-249-2118	3	8	4320-01-264-5703	4 15	
4710-01-249-2119	3	7	3020-01-264-5709	4 39	
4710-01-249-3576	3	6	4320-01-264-8408	4 83	
5360-01-250-0562	15	22	2910-01-264-8410	4 19	
5330-01-250-1390	3 14		4730-01-264-9460	4 91	
5330-01-250-1391	3 22		2910-01-265-0427	4 62	
3120-01-250-2116	11	9	5306-01-265-2059	15	6
	11	31	5306-01-265-2060	18	11

STOCK NUMBER	FIG.	ITEM	STOCK NUMBER	FIG.	ITEM
5306-01-265-2061	17	5	5305-01-266-9343	5	8
5306-01-265-2062	8 11			6	13
5310-01-265-2108	14	8	5307-01-266-9757	4 90	
5310-01-265-2109	6 16		5330-01-267-1231	8 28	
5310-01-265-2110	6 16		5306-01-267-1778	5	6
5310-01-265-2111	6 16		5360-01-267-2931	4	9
5310-01-265-2112	6 16		5340-01-267-2963	4 79	
5310-01-265-2113	6 16		5305-01-267-6310	4 51	
5310-01-265-2114	6 16		5305-01-267-6470	6	3
5310-01-265-2115	6 16		5360-01-267-6712	4 58	
5310-01-265-2116	6 16		5365-01-267-7498	4 18	
5310-01-265-2118	8 31		5305-01-267-8470	7	4
5310-01-265-2120	14	2	5330-01-267-9177	4 13	
5310-01-265-2121	2 22		5305-01-267-9222	7	2
5330-01-265-2127	4	2	5305-01-267-9223	4 85	
5340-01-265-2246	14	11	5305-01-267-9278	8 30	
3o40-01-265-2608	4 63		5305-01-267-9279	1 10	
2940-01-265-2668	5 11		5305-01-267-9281	17	8
5306-01-265-4697	19	35	5305-01-267-9286	14	1
5365-01-265-4921	6 16		5305-01-268-0138	1	5
5365-01-265-4922	6 16		5330-01-268-0159	4	8
5310-01-265-6689	17	6	5360-01-268-1019	4 14	
5340-01-265-6748	19	41	4730-01-268-2448	4	7
5340-01-265-6749	19	38	5340-01-268-2652	4 81	
5365-01-265-7125	9 14		5306-01-268-6796	6	14
5307-01-265-7140	18	15		8	15
				19	2
5365-01-265-7189	18	9	5365-01-268-6976	4 78	
	19	10	5306-01-268-8982	4 74	
5340-01-265-7263	15	10	5365-01-268-8993	4 12	
5340-01-265-7264	15	9	5305-01-268-9066	4	22
	19	8		4	67
	19	36		13	1
5340-01-265-7271	15	12	5365-01-268-9114	4 78	
5340-01-265-7276	8 26		5365-01-268-9115	4 78	
5365-01-266-4081	6 16		5365-01-268-9150	4 78	
5340-01-266-5116	15	11	5310-01-268-9750	4 78	
	19	40	5365-01-268-9151	4 78	
5340-01-266-5117	19	39			

STOCK NUMBER	FIG.	ITEM	STOCK NUMBER	FIG.	ITEM
5305-01-268-9758	4 82		5365-01-275-7181	4 78	
5365-01-269-0042	4 78		4730-01-280-4081	2	1
5365-01-269-0043	4 78		5365-01-289-4452	4 78	
5365-01-269-0044	4 78		3040-01-301-0405	11	15
5365-01-269-1049	4 78		4720-01-302-2944	BULK	BULK
5365-01-269-1050	4 78		2815-01-303-2505	8 24	
5365-01-269-1516	4 78		5330-01-305-0239	8 35	
4730-01-269-2313	1	14	5310-01-329-6975	8 13	
	15	5	4720-01-359-9633	BULK	BULK
5340-01-269-2709	16	15	2920-01-360-3314	10	2
5365-01-269-2713	4 78		5310-01-360-5911	14	18
5365-01-269-2714	4 78				4
5365-01-269-3783	4 78		5310-01-360-5961	6	2
5365-01-269-4446	4 78			12	3
5365-01-269-4448	4 78				5
5365-01-269-4449	4 78		2910-01-360-6368	5 10	
5365-01-269-4450	4 78		2910-01-360-6369	5	9
5365-01-269-4451	4 78		2815-01-360-6518	16	1
5365-01-269-4452	4 78		5310-01-360-7785	9	4
5365-01-269-4453	4 78		5330-01-360-9289	11	25
5340-01-269-5243	15	8	2910-01-361-0616	3	9
	19	9	5306-01-361-4586	12	7
5310-01-269-6280	4 78		5306-01-361-4588	19	27
5305-01-269-7534	4	1	5307-01-361-4612	9 10	
	13	10	5330-01-361-5287	4 89	
5305-01-270-0073	4 88		5310-01-361-5702	4 45	
5305-01-270-1704	4	23	5315-01-361-5705	16	13
	8	16	5365-01-361-5708	19	13
	19	21	5310-01-361-5718	4 47	
4730-01-270-6432	8	1	5310-01-361-5719	18	2
5310-01-270-8920	4 78			19	25
5310-01-271-1003	6 16		5310-01-361-5722	4	46
5310-01-271-7712	8 12			8	22
5340-01-272-4897	8	7		18	13
4030-01-272-7572	3	4	5310-01-361-5723	4	5
5306-01-272-9033	6	11	5365-01-361-5724	4 78	
	9	1	5365-01-361-5725	4 78	
5355-01-274-2456	13	6	5365-01-361-5726	4 78	

STOCK NUMBER	FIG.	ITEM	STOCK NUMBER	FIG.	ITEM
5365-01-361-6567	4 65		3020-01-362-5712	18	7
5365-01-361-6568	4 78		5310-01-362-5934	4	4
5365-01-361-6569	4 78		5310-01-362-5946	2 21	
5365-01-361-6570	4 78		3040-01-363-2360	4 30	
5306-01-361-6606	18	4	2910-01-363-2375	4 80	
5306-01-361-6607	8 25		2815-01-363-2382	16	14
5330-01-361-6633	6	8	4710-01-363-2490	1 11	
5365-01-361-6807	14	9	4710-01-363-3041	5	7
5306-01-361-6941	17	1	2910-01-363-3087	2	5
5330-01-361-6958	19	17	3020-01-363-3091	17	3
5310-01-361-7892	9	12	2930-01-363-3096	1	6
	12	6	2815-01-363-3098	9	5
5365-01-361-7899	4 65		4730-01-363-4324	19	43
5330-01-361-7912	4 75		2920-01-363-4364	11	8
5330-01-361-7913	13	9	2815-01-363-5159	15	25
5330-01-361-7914	13	5	4730-01-363-5161	1	8
5330-01-361-7915	14	6	2920-01-363-5170	11	26
5306-01-361-8359	1	1	2920-01-363-5178	11	28
5306-01-361-8360	4	94	2920-01-363-5181	11	22
	6	3	2815-01-364-1519	4 71	
	13	4	2929-01-364-1622	11	30
5306-01-361-8362	8 21		5305-01-364-3142	1 16	
5340-01-361-8366	19	15	3120-01-365-1738	16	18
5340-01-361-8372	8	2	5977-01-365-5584	11	10
5340-01-361-8415	11	29	2815-01-370-0208	7	3
5306-01-361-8433	1 19		5306-01-376-0730	1	4
5306-01-361-8434	2 20		5310-01-376-2850	12	25
5330-01-361-8991	4 61		5310-01-376-2860	11	6
5330-01-361-8992	13	3	5961-01-376-4821	12	24
5306-01-361-9044	6	5	5330-01-380-9110	2 12	
2815-01-362-1753	15	15	5330-01-380-9155	2 10	
4820-01-362-3399	8 14		5310-01-381-0737	12	4
2815-01-362-3546	19	33	5330-01-381-1934	6	2
5365-01-362-4406	4 65		5307-01-381-2792	4 50	
4730-01-362-5574	19	12	5307-01-381-2821	9	8
4730-01-362-5575	3 21		5305-01-381-2822	6 15	
2815-01-362-5668	15	24	2920-01-381-8693	11	13
2920-01-362-5694	12	28			

STOCK NUMBER	FIG.	ITEM	STOCK NUMBER	FIG.	ITEM
2815-01-381-9068	15	28	2815-01-406-1958	17	11
5305-01-381-9841	11	7	5330-01-413-3723	1 12	
5305-01-381-9939	12	33	5330-01-413-3724	5 17	
4720-01-382-2845	1	9	4730-01-414-0348	5	1
6115-01-382-3940	12	8	5330-01-414-0691	5 20	
5306-01-382-4201	12	21	5310-01-414-0741	11	3
5330-01-382-5915	2 18		5365-01-414-0745	4 65	
5339-01-382-5916	18	10	4730-01-414-1259	5 13	
	19	11	5340-01-414-2903	19	28
5330-01-383-6787	5 16		5365-01-414-3017	6	7
5330-01-383-8706	4 29		5306-01-414-3019	6 12	
5310-01-383-8817	12	13	5330-01-414-3640	5	2
5306-01-384-3476	18	14	5310-01-414-3661	11	2
2910-01-384-5937	4	3	5330-01-414-3709	5 12	
2920-01-385-5341	11	4	5330-01-414-3713	19	32
5306-01-388-2096	19	24	5310-01-414-4178	11	1
5365-01-388-5092	6 16		4710-01-414-9304	8 23	
2815-01-389-0315	19	44	2815-01-414-9430	14	4
5120-01-390-4740	20	9	5330-01-415-9615	4 54	
5120-01-390-7570	20	5	5330-01-416-2968	4 25	
5120-01-390-7574	20	7	5365-01-416-6508	19	31
5120-01-390-7577	20	2	5330-01-421-3133	15	20
5210-01-390-8341	20	10	5315-01-467-3123	4 70	
5120-01-390-8342	20	3	5306-01-467-4944	18	17
5120-01-390-8343	20	8	5365-01-467-5653	4 65	
5210-01-390-8344	20	11	5306-01-467-5689	18	1
5120-01-390-8345	20	1	5306-01-467-5694	18	18
5120-01-390-8346	20	6	4720-01-467-6675	8	3
5120-01-390-8347	20	4	6680-01-472-9712	6	1
5305-01-395-1862	11	27	2940-01-493-4533	5 10	
5977-01-395-1865	12	21	5340-01-556-2153	4 24	
2520-01-396-1397	11	21	5977-01-560-9133	12	22
5977-01-397-0407	11	14	6115-01-560-9146	12	8
2920-01-397-1166	12	29	5330-01-560-9319	2 14	
2920-01-397-5289	12	23	4710-01-560-9321	3	5
5340-01-397-5931	11	20	4710-01-560-9329	2	3
5120-01-398-3999	20	12	5305-01-560-9352	12	33

STOCK NUMBER	FIG.	ITEM
5360-01-560-9847	11	11
3120-01-560-9877	11	31
4710-01-560-9882	3	8
5360-01-560-9887	4 59	
5330-01-560-9889	4 69	
4710-01-560-9892	3	7
2915-01-561-0259	4	3
5340-01-561-0694	4 60	
5330-01-561-0828	4 16	
4330-01-561-0833	2 17	
4710-01-561-0841	3	6
5977-01-561-0860	11	12
	11	14
3120-01-561-0957	11	23
5306-01-561-1246	2	9
5306-01-561-1338	2	4
5330-01-561-1363	2 16	
5310-01-561-1396	2 15	
5310-12-136-5792	12	16
5305-12-185-6163	12	17
4820-66-128-5601	8 10	

END OF WORK PACKAGE

THE METRIC SYSTEM AND EQUIVALENTS

METRIC SYSTEM AND EQUIVALENTS

LINEAR MEASURE

1 Centimeter = 10 Millimeter = 0.01 Meters = 0.3937 Inches

1 Meter = 100 Centimeters = 1000 Millimeters = 39.37 Inches

1 kilometer = 1000 Meters = 0.621 Miles

WEIGHTS

1 Gram = 0.0⁰¹ Kilograms = 1000 Milligrams = 0.035 Ounces

1 Kilogram = 100 Grams = 2.2 lb.1 Cu. Meter = 1,000,000

1 Metric Ton = 1000 Kilograms = 1 Megagram = 1.1 Short Tons

LIQUID MEASURE

1 Millimeter = 0.001 Liters = 0.0338 Fluid Ounces

1 Liter = 1000 Millimeters = 32.82 Fluid Ounces

SQUARE MEASURE

1 Sq. Centimeter = 100 Sq. Millimeter = 0.155 Sq. Inches

1 Sq. Meter = 10,000 Sq. Centimeters = 10.76 Sq. Inches

1 Sq. Kilometer = 1,000,000 Sq. Meters = 0.386 Sq. Miles

CUBIC MEASURE

1 Cu. Centimeter = 1000 Cu. Millimeters = 0.06 Cu. Inches

1 Cu. Centimeters = 35.31 Cu. Feet

TEMPERATURE

5/9 (°F - 32) = °C

212° Fahrenheit is equivalent to 100° Celsius

90° Fahrenheit is equivalent to 32.2° Celsius

32° Fahrenheit is equivalent to 0° Celsius

9/5 °C + 32 = °F

APPROXIMATE CONVERSION FACTORS

TO CHANGE	TO	MULTIPLY BY
Inches	Centimeters	2.540
Feet	Meters	0.305
Yards	Meters	0.914
Miles	Kilometers	1.609
Square Inches	Square Centimeters	6.451
Square Feet	Square Meters	0.093
Square Yards	Square Meters	0.836
Square Miles	Square Kilometers	2.590
Acres	Square Hectometers	0.405
Cubic Feet	Cubic Meters	0.028
Cubic Yards	Cubic Meters	0.765
Fluid Ounces	Milliliters	29.573
Pints	Liters	0.473
Quarts	Liters	0.946
Gallons	Liters	3.785
Ounces	Grams	28.349
Pounds	Kilograms	0.454
Short Tons	Metric Tons	0.907
Pound-Feet	Newton-Meters	1.356
Pounds per Square Inch	Kilo pascals	6.895
Miles per Gallon	Kilometers per Liter	0.425
Miles per Hour	Kilometers per Hour	1.609

TO CHANGE	TO	DIVIDE BY
Centimeters	Inches	2.540
Meters	Feet	0.305
Meters	Yards	0.914
Kilometers	Miles	1.609
Square Centimeters	Square Inches	6.451
Square Meters	Square Feet	0.093
Square Meters	Square Yards	0.836
Square Kilometers	Square Miles	2.590
Square Hectometers	Acres	0.405
Cubic Meters	Cubic Feet	0.028
Cubic Meters	Cubic Yards	0.765
Milliliters	Fluid Ounces	29.573
Liters	Pints	0.473
Liters	Quarts	0.946
Liters-Meters	Gallons	3.785
Grams	Ounces	28.349
Kilograms	Pounds	0.454
Metric Tons	Short Tons	0.907
Newton-Meters	Pound-Feet	1.356
Kilo pascals	Pounds per Square Inch	6.895
Kilometers per Liter	Miles per Gallon	0.425
Kilometers per Hour	Miles per Hour	1.609

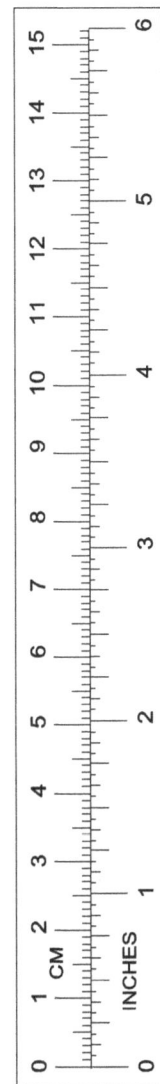

www.ingramcontent.com/pod-product-compliance
Lightning Source LLC
Chambersburg PA
CBHW080416030426
42335CB00020B/2464